RESOURCE ALLOCATION IN UPLINK OFDMA WIRELESS SYSTEMS

RESOURCE ALLOCATION IN UPLINK OFDMA WIRELESS SYSTEMS

Optimal Solutions and Practical Implementations

ELIAS E. YAACOUB

QU Wireless Innovations Center

ZAHER DAWY

American University of Beirut

 John B. Anderson, *Series Editor*

IEEE PRESS

A JOHN WILEY & SONS, INC., PUBLICATION

Published by John Wiley & Sons, Inc., Hoboken, New Jersey
Published simultaneously in Canada

For general information on our other products and services or for technical support, please contact our Customer Care Department within the United States at (800) 762-2974, outside the United States at (317) 572-3993 or fax (317) 572-4002.

Wiley also publishes its books in a variety of electronic formats. Some content that appears in print may not be available in electronic formats. For more information about Wiley products, visit our web site at www.wiley.com.

Library of Congress Catalog Number: 58-9935

Resource Allocation in Uplink OFDMA Wireless Systems / Elias E. Yaacoub and Zaher H. Dawy.
    p. cm.–(Wiley IEEE Series on Digital and Mobile Communication)
"Wiley-Interscience."
Includes bibliographical references and index.
ISBN 0-471-48348-6 (pbk.)
1. Surveys–Methodology. 2. Social sciences–Research–Statistical methods.
I. Groves, Robert M. II. Series.
HA31.2.S873 2007
001.4'33–dc22   2004044064

Printed in the United States of America.
10 9 8 7 6 5 4 3 2 1

To Therese and Maria Theresa
Elias Yaacoub
To Sanaa, Hassan, and Noura
Zaher Dawy

CONTENTS

PREFACE

In the era of broadband wireless connectivity, users are expecting ubiquitous and seamless access to a wide range of bandwidth demanding services. Orthogonal frequency division multiple access (OFDMA) has been selected as the accessing scheme for state-of-the-art wireless communication systems. The high data rates and low latency expected from current and next generation broadband wireless access systems, such as 3GPP long-term evolution (LTE) and WiMAX (IEEE 802.16e/m), necessitate optimized and dynamic allocation of the available radio resources. In addition, the increase in demand for delay-sensitive applications with bidirectional data rate requirements such as wireless gaming, video telephony, and voice-over-IP, mandates the need for optimized uplink resource allocation algorithms.

This book treats the area of resource allocation in OFDMA wireless systems, with a focus on the uplink direction. However, the downlink direction is widely discussed, and most of the presented techniques and insights apply to the downlink. The book investigates the problems of single cell resource allocation, multicell resource allocation, centralized resource allocation, and distributed resource allocation, with/without collaboration among base stations and with/without cooperation among mobile users. Resource allocation in OFDMA wireless systems will constitute an important topic for several years to come, especially in the context of distributed and multicell scenarios. Algorithms and techniques that lead to efficient performance results, while at the same time being simple enough for practical implementation, will be of particular importance. The presented techniques in this book touch as well upon applications in cognitive radio networks, distributed base stations, heterogeneous wireless networks, and femtocells.

This book is addressed to students, professors, and researchers whose research is in the area of wireless communications with focus on resource allocation, interference mitigation, radio resource management, heterogeneous networks, and applications of game theory and convex optimization in wireless communications. The book has classroom adoption possibilities. It could be considered as a valuable supplemental reading in courses on wireless communications, cellular technologies, resource management, and network optimization. It could also serve as a supplemental reading for convex optimization and game theory courses since it presents several applications of important concepts studied in these courses in a wireless communications framework. In addition, the book is addressed to research and development engineers working in the telecom industry on next generation wireless cellular technologies. In fact, the algorithms and techniques presented in the book can be customized for possible implementation in current- and next-generation LTE and WiMAX base stations. The book is also useful for telecommunications operators and vendors, service providers, consultants, research centers, and standardization bodies working on the next-generation of cellular technologies, like LTE-Advanced and WiMAX IEEE 802.16m.

A brief overview of the book organization is as follows. Chapters 1 and 2 are of introductory nature: Chapter 1 is a general introduction for the book whereas Chapter 2 presents a high-level summary of downlink OFDMA resource allocation. Centralized scheduling in a single cell scenario is studied in Chapters 3–7. Theoretical techniques based on optimization problems formulation and derivation of the optimal solutions are presented in Chapters 3–5. The reader interested in practical suboptimal schemes can skip these chapters and start reading from Chapter 6. Distributed resource allocation within a single cell scenario is investigated in Chapters 8–10, with Chapter 8 representing an intermediate step between fully centralized and fully distributed resource allocation. Multicell resource allocation is studied in Chapters 11 and 12. Practical implementation aspects of the proposed techniques and their relation to the LTE and WiMAX standards are discussed in Chapter 13. Finally, Chapter 14 discusses open research directions worthy of further investigation.

ACKNOWLEDGEMENTS

The authors would like to thank Ahmad El-Hajj for his valuable contributions and suggestions, especially during the preparation of Chapter 3. The authors are also grateful for the fruitful discussions with Profs. Mohamad Adnan Al-Alaoui, Karim Kabalan, and Ibrahim Abou Faycal from AUB, in addition to Prof. Mohamed-Slim Alouini from KAUST. The authors appreciate the encouragement of Dean Ibrahim Hajj from AUB and Dr. Adnan Abu-Dayya from QUWIC, during the preparation of this book. In addition, the authors would like to express their gratitude toward the reviewers, who contributed to shaping the final version of the book, and the team at John Wiley & Sons Inc. for their support.

The authors wish to thank the following organizations and funding sources for their support during the preparation of this book: the American University of Beirut (AUB), the AUB University Research Board, Dar Al-Handasah (Shair and Partners) Research Fund, Rathmann Family Foundation Fund, QU Wireless Innovations Center (QUWIC), Qatar National Research Fund (QNRF), and Qatar Telecom (QTel).

ACRONYMS

3GPP	Third Generation Partnership Project
AMC	Adaptive Modulation and Coding
AP	Access Point
BE	Best Effort
BER	Bit Error Rate
BPSK	Binary Phase Shift Keying
BS	Base Station
CBR	Constant Bit Rate
CCD	Central Controlling Device
CCU	Central Control Unit
CDMA	Code Division Multiple Access
CINR	Channel Gain to Interference Plus Noise Ratio
CNR	Channel Gain to Noise Ratio
CP	Cyclic Prefix
CQI	Channel Quality Indicator
CR	Cognitive Radio
CSI	Channel State Information
DAS	Distributed Antenna System
DBS	Distributed Base Station
DFT	Discrete Fourier Transform
DSL	Digital Subscriber Line
FDD	Frequency Division Duplex
FDMA	Frequency Division Multiple Access
FFT	Fast Fourier Transform
GSM	Global System for Mobile Communications
HII	High Interference Indicator
IDFT	Inverse Discrete Fourier Transform
IFDMA	Interleaved Frequency Division Multiple Access
IFFT	Inverse Fast Fourier Transform
IID	Independent and Identically Distributed
KKT	Karush–Kuhn–Tucker
LFDMA	Localized Frequency Division Multiple Access
LTE	Long-Term Evolution
MAC	Medium Access Control
MCS	Modulation and Coding Scheme
MIMO	Multiple Input Multiple Output
MRC	Maximum Ratio Combining
NBP	Nash Bargaining Problem

NBS	Nash Bargaining Solution
NE	Nash Equilibrium
OFDM	Orthogonal Frequency Division Multiplexing
OFDMA	Orthogonal Frequency Division Multiple Access
OI	Overload Indicator
PDF	Probability Density Function
PF	Proportional Fair
PFF	Proportional Fair in Frequency
PFT	Proportional Fair in Time
PFTF	Proportional Fair in Time and Frequency
PHY	Physical Layer
QAM	Quadrature Amplitude Modulation
QoS	Quality of Service
QPSK	Quadrature Phase Shift Keying
RB	Resource Block
RR	Round Robin
RRC	Radio Resource Controller
RRH	Remote Radio Head
RRM	Radio Resource Management
SC	Selection Combining
SCFDMA	Single Carrier Frequency Division Multiple Access
SINR	Signal to Interference Plus Noise Ratio
SNR	Signal-to-Noise Ratio
SRS	Sounding Reference Signal
SS	Subscriber Station
TDD	Time Division Duplex
TDMA	Time Division Multiple Access
TTI	Transmission Time Interval
UE	User Equipment
UMTS	Universal Mobile Telecommunications System
UTRAN	UMTS Terrestrial Radio Access Network
VBR	Variable Bit Rate
WiMAX	Wireless Interoperability for Microwave Access
WLAN	Wireless Local Area Network

CHAPTER 1

INTRODUCTION

1.1 EVOLUTION OF WIRELESS COMMUNICATION SYSTEMS

Wireless cellular technologies are continuously evolving to meet the increasing demands for high data rate mobile services. The design of cellular wireless communications with spectrum reuse was first developed in the 1960s. The first generation of cellular systems was based on analog frequency modulation (FM) and frequency division multiple access (FDMA) with voice traffic as the only service provided [1]. After the world's first cellular system was deployed in Japan by Nippon Telephone and Telegraph (NTT) in 1979, the Advanced Mobile Phone System (AMPS) in the United States and the European Total Access Cellular System (ETACS) in Europe followed in the 1980s [1, 2]. The design of the second generation of cellular systems moved from analog to digital communications due to the numerous advantages of the latter: support for efficient voice compression, advanced digital signal processing techniques at the transmitter and receiver, error correction coding, in addition to cheaper and faster components requiring less transmit power capabilities.

The second generation Global System for Mobile (GSM) Communications is currently the most widely deployed cellular system with more than three billion subscribers in more than 200 countries [3]. GSM is based on a hybrid time division/frequency division multiple accessing (TDMA/FDMA) air interface. The allocated spectrum for a given GSM network is divided into multiple 200 KHz frequency channels that are distributed among cells in the network based on a frequency plan in order to control the level of intercell interference in the network. In GSM, time is divided into eight time slots where each active user is allocated a time slot on a selected frequency channel. The proliferation of the Internet and the continuously escalating market demand for mobile data services necessitated the evolution of GSM by developing new enhancement technologies such as General Packet Radio Service (GPRS), enhanced data rates for GSM evolution (EDGE), and evolved EDGE that are capable of supporting packet switched mobile services with peak data rates of 128 kbps, 384 kbps, and 1 Mbps, respectively. The enhanced data rates offered by these technologies are based on various techniques that include adaptive modulation and coding, allocation of multiple time slots per user, in addition to dual-carrier downlink and dual-antenna terminals in evolved EDGE.

Resource Allocation in Uplink OFDMA Wireless Systems: Optimal Solutions and Practical Implementations, Elias E. Yaacoub and Zaher Dawy.
© 2012 by the Institute of Electrical and Electronics Engineers, Inc. Published 2012 by John Wiley & Sons, Inc.

In order to meet the forecasted growth of cellular subscribers and the need for faster and more reliable data services, the third generation Universal Mobile Telecommunication System (UMTS) was developed and standardized in 1999, based on a code division multiple accessing (CDMA) air interface. In a UMTS network, the frequency reuse factor is one, and thus all users in the network share the same frequency band at the same time by using different spreading/scrambling codes in order to limit the level of intracell and intercell interference in the network. The Wideband CDMA (WCDMA) mode of UMTS is currently being widely deployed by cellular operators all over the world with more than 500 million subscribers in more than 130 countries [4]. WCDMA provides higher spectral efficiency than GSM/GPRS/EDGE, with peak downlink data rates theoretically up to 2 Mbps in 3GPP Release'99, and beyond 10 Mbps in 3GPP Release 5. Practical bit rates are up to 384 kbps initially, and beyond 2 Mbps with Release 5 [5]. High Speed Packet Access (HSPA) refers to two enhancement technologies for UMTS networks: High Speed Downlink Packet Access (HSDPA) with peak bit rates up to 14.4 Mbps in the downlink and High Speed Uplink Packet Access (HSUPA) with peak bit rates up to 5.7 Mbps in the uplink. HSPA+, referred to as enhanced HSPA, can provide peak bit rates up to 84 Mbps in the downlink and 21 Mbps in the uplink. The high data rates achieved by HSPA and HSPA+ are based on various advanced features that include intelligent fast scheduling at the base station (BS), adaptive modulation and coding with high-order MQAM modulation schemes, fast data retransmission over the air interface, and multiple antenna techniques such as spatial multiplexing in addition to open loop and closed loop transmit diversity.

The 3GPP Long-Term Evolution (LTE) has been standardized as the next generation of cellular technologies following UMTS/HSPA+ on the evolution track. The LTE standard is based on an orthogonal frequency division multiple accessing (OFDMA) air interface and is capable of providing notably high peak data rates up to 75 Mbps in the uplink and 300 Mbps in the downlink. At the same time, the Wireless Interoperability for Microwave Access (WiMAX) wireless technology is being further developed as an all-IP network with an OFDMA air interface, to provide high data rate broadband wireless mobile access [6]. WiMAX evolved from a last mile wireless access technology to a broadband wireless access technology, with nomadic and mobile connectivity as standardized in IEEE 802.16d and IEEE 802.16e, respectively. Currently, standardization activities resulted in developing the technical specifications for the IMT-Advanced technologies LTE-Advanced and WiMAX IEEE 802.16m with peak bit rate capabilities exceeding 1 Gbps in the downlink.

1.2   ORTHOGONAL FREQUENCY DIVISION MULTIPLE ACCESS

Orthogonal frequency division multiplexing belongs to the family of multicarrier modulation schemes. It is based on dividing the transmitted bitstream into multiple substreams and sending these over different orthogonal subcarriers, also called subchannels. In OFDM, the number of subcarriers is selected such that each subcarrier

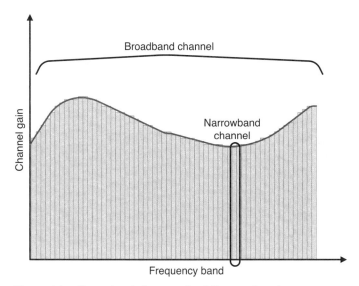

Figure 1.1   Channel variation over the different subcarriers.

has a bandwidth less than the coherence bandwidth of the channel as illustrated in Fig. 1.1, in order for the subcarriers to experience relatively flat fading and, thus, avoid intersymbol interference. The subcarriers in OFDM are not required to be contiguous as in Fig. 1.1. Thus, a large continuous block of spectrum is not needed for high rate multicarrier communications, and several contiguous blocks of smaller size can be used instead. This provides flexibility in spectrum allocation and spectrum management.

Multicarrier modulation schemes using overlapping but orthogonal subcarriers were investigated since the 1960s [7, 8]. However, the use of such schemes at the time was practically difficult due to the large number of filters and modulators required [9]. A major reduction in the implementation complexity of OFDM transmitters and receivers was achieved by using discrete Fourier transform (DFT) operations to modulate and demodulate OFDM signals [10]. With the DFT implementation, frequency division is achieved by baseband processing instead of bandpass filtering. An OFDM transmitter and receiver implementation using fast Fourier transform (FFT) blocks is shown in Fig. 1.2. OFDM modulation is used in cable access networks, such as Asymmetrical Digital Subscriber Lines (ADSL) and Hybrid Digital Subscriber Lines (HDSL) [11–13], in addition to several wireless communication systems, such as Wireless LANs (WLANs) [14], Digital Video Broadcasting (DVB) [15], LTE, and WiMAX. A historical overview of OFDM can be found in Ref. [16].

The main advantages of OFDM include robustness against multipath fading, exploitation of frequency diversity, facilitation of advanced multiple input multiple output (MIMO) techniques, in addition to adaptive loading per subcarrier and efficient multiple accessing. The basic idea of adaptive loading is to vary the data rate and power assigned to each subcarrier depending on its channel gain. Hence, the power and rate associated with each subcarrier can be optimized to maximize the rate for a

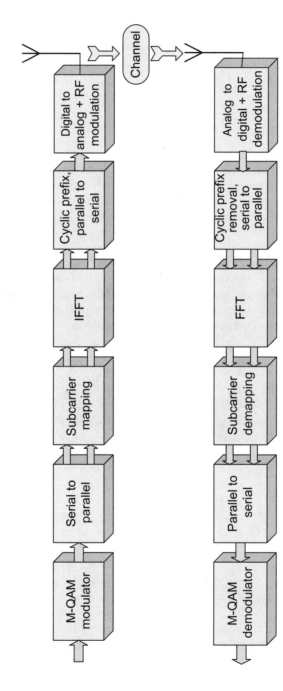

Figure 1.2 OFDM with IFFT/FFT implementation.

given maximum transmit power or to minimize the transmit power for a given target rate. This can be achieved by using a variable-rate variable-power modulation scheme like MQAM. The power loading on the different subcarriers was first investigated in Ref. [17]. Orthogonal frequency division multiple access (OFDMA) is an extension of OFDM based on dividing the subcarriers among users in order to exploit multiuser diversity gains, which made it an attractive choice for cellular broadband wireless access systems such as LTE and WiMAX. A major challenge to exploit the advantages of OFDMA in wireless broadband communication systems is the efficient joint allocation of subcarriers and powers among users in the uplink and downlink in order to meet target quality of service objectives such as target bit rate, latency, and/or fairness constraints. This requires the design of dynamic and optimized resource allocation schemes that can adapt based on varying channel and interference conditions and that can be enhanced to exploit cooperation opportunities among base stations and users while maintaining relatively low complexity suitable for practical implementation.

To this end, this book presents a comprehensive study of resource allocation and scheduling techniques in OFDMA wireless networks (the terms "resource allocation" and "scheduling" are used interchangeably throughout the book). Although the investigation of both the uplink and downlink directions is of equal importance, the main focus in this book will be on the uplink. In fact, the increase in demand for mobile applications with bidirectional rate and delay requirements has mandated the need for efficient uplink scheduling algorithms in OFDMA wireless systems. This book treats the area of uplink OFDMA resource allocation from various perspectives that include the following: centralized and distributed, instantaneous and ergodic, optimal and suboptimal, single cell and multicell, cooperative and noncooperative, in addition to different combinations of these variants.

1.3 ORGANIZATION OF THIS BOOK

An overview of the book outline is shown in Fig. 1.3. Chapter 2 is a background chapter that presents a survey of downlink resource allocation and scheduling techniques in OFDMA wireless networks. In Chapter 3, the problem of ergodic sum-rate maximization with continuous rates in OFDMA uplink is formulated and solved. Power and rate constraints are considered, and the achievable rate region is investigated. Chapter 4 presents the formulation and solution of the ergodic sum-rate maximization problem with discrete rates. This represents an extension of Chapter 3 to a scenario where a limited number of achievable rates are available due to a predefined set of modulation and coding schemes. In Chapter 5, the solutions of Chapters 3 and 4 are extended to general utility maximization, where the utility of a user is a function of its rate. Chapter 6 presents suboptimal scheduling algorithms that overcome the practical limitations of the optimal solutions derived in Chapters 3–5. The suboptimal algorithms are shown to achieve a close performance to the optimal solutions, with a considerably reduced complexity. In Chapter 7, the suboptimal algorithms are applied with various utility functions. An emphasis is given to utilities ensuring proportional fairness in order to provide fair access to the resources among the users.

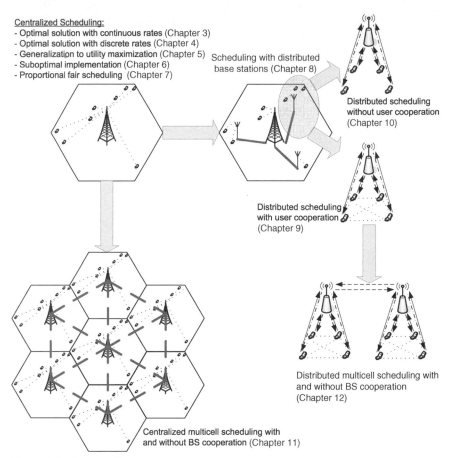

Figure 1.3　Book outline.

Chapters 3–7 deal with fully centralized scheduling. Chapter 8 discusses the implementation of the algorithms in a distributed base station scenario. This chapter constitutes an intermediate step between centralized scheduling where full control is given to the BS, and distributed scheduling where users take part in scheduling decisions. Chapter 9 treats the scenario of distributed resource allocation with user cooperation. Users are assumed to be able to successfully exchange information. Hence, this represents a scenario with a limited coverage area, for example, the area covered by a single remote radio head (RRH) in a distributed base station system. The users implement the algorithms of Chapters 6 and 7 in a distributed way, and achieve results close to centralized scheduling with a limited amount of exchanged information. Chapter 10 consists of investigating distributed resource allocation without user collaboration. The proposed approach is based on channel sensing and probabilistic transmission. It consists of a scheduling phase followed by a transmission phase. The approach leads to complete avoidance of collisions during transmission, while reducing the collision probability during the scheduling phase.

Chapter 11 treats the problem of centralized scheduling in a multicell scenario. The resource allocation problem is formulated as a pricing game between BSs and the convergence to a Nash equilibrium is shown. Enhancements in the presence of a central controller are demonstrated. Suboptimal scheduling techniques are presented to overcome the limitations of the game theoretic model. The suboptimal techniques rely on pricing-based power control in the presence and absence of BS collaboration, in addition to probabilistic scheduling in the latter case. Chapter 12 investigates the problem of distributed scheduling in a multicell scenario. Techniques derived in Chapter 11 are applied to the distributed scheduling scenario presented in Chapter 9. In addition, a transparent pricing scheme is presented in the case with BS collaboration, where the prices are completely oblivious to the users.

Chapter 13 presents an overview of resource allocation in the state-of-the-art wireless systems LTE and WiMAX, and describes the applicability of the presented resource allocation techniques in these systems. Finally, open research directions in the area of OFDMA resource allocation are outlined in Chapter 14.

CHAPTER 2

BACKGROUND ON DOWNLINK RESOURCE ALLOCATION IN OFDMA WIRELESS NETWORKS

This chapter is a background chapter that presents a survey of downlink resource allocation and scheduling techniques in OFDMA wireless networks. In the subsequent chapters of this book, background on relevant resource allocation techniques for both OFDMA uplink and downlink will be presented and analyzed. However, the main focus will be on the uplink, while noting that several of these techniques can be applied to the downlink direction also.

In this chapter, the main aspects of downlink resource allocation are summarized, and the interested reader can find an overview of various OFDMA downlink resource allocation and scheduling scenarios: centralized and distributed, instantaneous and ergodic, optimal and suboptimal, single cell and multicell, cooperative and noncooperative, in addition to different combinations of these variants. Applications to the LTE system are presented, and the additional constraints imposed by the LTE standard are outlined.

The chapter is organized as follows. Centralized scheduling techniques within a single cell are described in Section 2.1. Scenarios with distributed scheduling are presented in Section 2.2. Section 2.3 surveys scheduling techniques in a multicell scenario where intercell interference represents a major challenge to be addressed. The chapter is summarized in Section 2.4.

2.1 CENTRALIZED SINGLE CELL SCHEDULING

In centralized scheduling, both in the downlink and uplink, the base station (BS) is responsible for the scheduling process. Decisions are made at the BS and communicated to mobile users. Centralized single cell downlink resource allocation in cellular OFDMA systems has been widely investigated in the literature [18–27]. Topics investigated include sum-rate maximization, general utility maximization, achieving a desired quality of service (QoS) with minimum power, and ergodic sum-rate maximization. In sum-rate maximization, which could be considered as an extension of

Resource Allocation in Uplink OFDMA Wireless Systems: Optimal Solutions and Practical Implementations, Elias E. Yaacoub and Zaher Dawy.
© 2012 by the Institute of Electrical and Electronics Engineers, Inc. Published 2012 by John Wiley & Sons, Inc.

the centralized maximum C/I scheduling scheme [28], it was shown that the optimal solution is to separate subcarrier allocation from power allocation: each subcarrier is allocated to the user with the best channel condition and power is allocated by water-filling over the subcarriers [19, 20, 24]. To provide more fairness, utility maximization is investigated, where the utility could be a function ensuring more fairness than using only the rate [19, 20, 24]. The logarithm of the rate, for example, is shown in Refs [19, 20] to achieve proportional fairness (PF) [29]. These works make use of the classical Shannon capacity (or some formula derived from it) to model the rate. Instead of trying to achieve the maximum capacity, another objective is to achieve a desired QoS while minimizing the required transmit power. This is done by bit loading over the allocated subcarriers [18, 21, 25, 30]. Most of the existing work focuses on channel aware resource allocation assuming the channel state information (CSI) is known at the scheduler, and that there is always data to transmit. However, some existing work deals with channel-aware queue-aware algorithms where the buffer lengths and the queue states of each user are taken into account [23].

In centralized instantaneous scheduling, a certain utility (e.g., sum-rate) is maximized at each scheduling instant. The BS takes advantage of the random variations of the channel states of the various users over the OFDMA subcarriers in order to perform efficient resource allocation. This way, the BS takes advantage of the frequency and multiuser diversity dimensions [18–20]. Ergodic scheduling uses an additional degree of freedom. With ergodic scheduling, the time dimension is used in addition to the frequency and multiuser diversity dimensions [26, 27, 31], and hence the interest is in long-term performance. In both the instantaneous and ergodic scheduling scenarios, the utility maximization is used either using continuous rates (Shannon capacity), or a discrete set of rates corresponding to a finite number of modulation and coding schemes (MCS).

The scheduling problem in the uplink is more challenging than that in the downlink due to the distributed power constraint: in the downlink, the power has a centralized nature because power allocation is done at a central entity, the BS, which is the single power source in a single cell. In the uplink, each user is a power source by itself. Hence, the power has a distributive nature and should be considered on a per-user basis.

The optimal solution in instantaneous sum-rate maximization in OFDMA uplink requires the computation of Lagrangian parameters enforcing the power constraint for each user being scheduled. In the downlink, a single Lagrangian parameter is needed since only one power constraint is available, the power constraint for the BS [26]. These parameters are generally computed using subgradient techniques [27, 32]. The convergence of subgradient iterations for a large number of users at each scheduling interval leads to a considerable complexity. This complexity can be reduced significantly by resorting to ergodic scheduling and taking advantage of the time dimension as an additional degree of freedom in the scheduling process.

Ergodic weighted sum-rate maximization in downlink OFDMA systems is considered in Refs [26, 27], subject to maximum power constraints. Ergodic weighted sum-rate maximization is also considered in Ref. [33] in the context of an ad hoc cognitive radio (CR) network. In Ref. [33], additional minimum rate constraints are added, in order to ensure fairness for users by allowing them to achieve a minimum target rate. The solution approach in Refs [26, 27, 33] consists of transforming the weighted sum-rate maximization problem into a convex utility maximization problem, formulating and solving the dual problem, and proving that the duality gap is zero. Hence, the solution of the dual problem corresponds to the solution of the primal problem [34]. To enforce the maximum transmit power constraint, in addition to the minimum rate constraint in Refs [27, 33], the Lagrangian parameters are computed via subgradient iterations as described in Ref. [35].

To perform these iterations with ergodic scheduling, if the fading probability distribution function (pdf) is known, enough samples can be generated offline to be used in the iterations of these Lagrangian parameters. The obtained Lagrangians can then be used in instantaneous resource allocation without recomputation as long as the fading pdf remains unchanged. In case the pdf is unknown and offline training is not possible, convergence can be obtained online by using running averages over the instantaneous fading realizations. In this case, it is not required that the BS knows the fading pdf, as long as it is aware of the instantaneous fading realizations. Such an online training approach is shown in Ref. [36] to lead to the same solution as the offline case in the context of ad hoc peer-to-peer cognitive radio networks. However, in instantaneous utility maximization, the Lagrangians have to be computed at each scheduling instant. This requires subgradient iterations in order to compute these parameters at every transmission time interval (TTI). Ergodic utility maximization allows to avoid this overhead. It should be noted that in both ergodic and instantaneous scheduling, the rate, subcarrier, and power allocations are performed at every fading state. The difference is in computing the Lagrangians: With ergodic scheduling, the optimal rate, subcarrier, and power allocations are computed at each fading state using the current Lagrangian parameters, acting as power prices in order to regulate the transmit power; at a slower timescale, the power prices are adjusted to meet the average power constraints, similarly to the approach described in Ref. [37]. However, with instantaneous scheduling, the power prices are adjusted at every fading state.

2.1.1 Continuous Versus Discrete Rates

The Shannon capacity formula for Gaussian channels, $\log_2(1 + \text{SNR})$, where SNR denotes the signal-to-noise ratio, is widely used in the literature, for example [19, 20, 24, 26, 27, 33]. This capacity expression is based on the assumption of infinite length codewords generated according to a normal distribution [38]. However, in a practical system, only a discrete set of rates are achievable due to a fixed number of MCS.

Most of the algorithms in the literature that use discrete rates treat the problem of the transmit power minimization given per user minimum rate constraints [18, 25, 30, 39]. This problem is the dual formulation of the sum-rate maximization problem, which consists of maximizing the sum-rate given a maximum power constraint. In Ref. [40], an algorithm proposed for power minimization is extended to the case of sum-rate maximization. The main focus in the literature is on downlink resource allocation [18, 25, 39, 40] with a few papers treating the uplink problem [30], where the power minimization problem is considered. These algorithms follow a three-steps approach to resource allocation with discrete rates: estimating the number of subcarriers to be allocated to each user, selecting and allocating the appropriate subcarriers, and using bit loading to allocate the power on these subcarriers. A comprehensive survey of existing algorithms is presented in Ref. [9]. Selecting the best subalgorithm in the literature for each of the three steps, a new algorithm is developed in Ref. [9], where the focus is on the downlink power minimization problem. However, indications on extending the algorithm to the sum-rate maximization case are discussed based on the approach of Ref. [40].

The algorithms in Refs [9, 18, 25, 30, 39, 40] correspond to instantaneous scheduling with discrete rates. The problem of ergodic sum-rate maximization with discrete rates is investigated in Ref. [26] for the downlink in addition to the continuous rates scenario. Due to the discontinuity of the maximization function, the problem becomes harder to solve due to the loss of convexity. However, it is shown in Ref. [26] that the maximization problem with discrete rates is quasiconcave and the optimal dual solution can be reached with zero duality gap as in the continuous rates scenario.

2.1.2 Optimal Versus Suboptimal Scheduling

Although the ergodic sum-rate maximization solutions are simpler to implement than the optimal instantaneous sum-rate maximization solutions, they require an initiation phase to compute the Lagrangians via iterative subgradient iterations, and a tracking of the channel probability density function to repeat the calculations when necessary. The computational load increases at the BS with the number of users, since the optimal solution necessitates that the transmit power dedicated to each user on each allocated subcarrier be computed every TTI, because the optimal solution is determined by water-filling. Hence, the investigation of suboptimal algorithms that achieve a good performance has received considerable attention in the literature [19, 20, 24, 25, 40].

It should be noted that when the number of subcarriers increases, the scheduling complexity increases, even with linear complexity algorithms. In practice, the signaling load will also increase with the number of subcarriers. Hence, resource allocation is not performed on a subcarrier by subcarrier basis in state-of-the-art OFDMA-based systems. Instead, subcarriers are assembled into groups of consecutive subcarriers, called subchannels in WiMAX or resource blocks (RB) in LTE for example, and allocation is performed on an RB basis. In LTE, the available spectrum is divided into RBs consisting of 12 adjacent subcarriers [41]. In WiMAX, 192 data OFDM subcarriers are distributed in 16 subchannels of 12 subcarriers each. Each subchannel is made of four groups of three adjacent subcarriers each [42]. An example of grouping subcarriers into blocks is shown in Fig. 2.1.

As stated previously, utility maximization is widely investigated [19, 20]. In addition to sum-rate maximization, where the utility of a user is its achieved rate, proportional fair scheduling is widely investigated in the literature, where the utility is the logarithm of user rate. The importance of proportional fair scheduling in providing fairness is explained in Ref. [43] by resorting to a game theoretical interpretation. The Nash bargaining problem (NBP) [44] is a well-known scenario in game theory. Players in the NBP negotiate to maximize their payoffs, given a set of shared resources. The optimal solution of the NBP, the Nash bargaining solution (NBS), consists of distributing the resources in a way to maximize the product of the payoffs [45]. It was shown that PF scheduling is equivalent to the implementation of the NBS in the

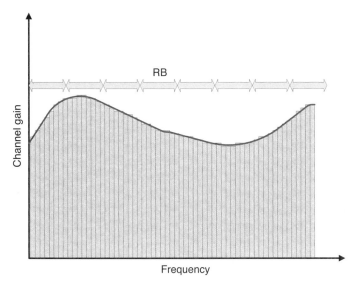

Figure 2.1   Adjacent subcarriers grouped into blocks.

resource allocation of wireless communication systems, the payoff of each user being its rate [43, 46]. PF scheduling is widely investigated in the literature, mainly in the framework of centralized resource allocation [19, 29]. With OFDMA adopted as the accessing scheme of next generation cellular systems, for example, 3GPP LTE and mobile WiMAX (IEEE 802.16e), several applications of PF to OFDMA have been studied [47–49].

2.2   DISTRIBUTED SCHEDULING

In centralized scheduling, resource allocation decisions are made at a central entity, the base station. In current and future broadband wireless access systems, users are expecting ubiquitous and seamless access to a variety of bandwidth demanding services. Mobile devices capable of supporting multiple standards are becoming more common in the market. Current research is not only ongoing on enhancing scheduling techniques within a given network, but also on optimizing the resource allocation in heterogeneous networks. This involves selecting the best network to serve a mobile user, among several networks with different access technologies such as GSM/EDGE, UMTS/HSPA, WiMAX, and WLAN [50–52].

Conversely to centralized resource allocation, mobile devices have more autonomy in making transmission decisions in distributed schemes. Distributed scheduling is usually studied in the context of ad hoc networks, relay-based networks, and sensor networks [53–55]. Distributed channel allocation schemes for wireless local area networks (WLANs) are also an active topic of current research [56, 57]. In addition, CR networks have gained increasing importance, and the problem of resource allocation in CR networks is being widely investigated [58–63]. CR, ad hoc, and sensor networks are distributed in nature. However, distributed resource allocation has also been implemented in infrastructure-based networks where users are connected to a central BS [64, 65].

In OFDMA distributed scheduling in the presence of a certain infrastructure, distributed antennas are placed throughout the cell area while being connected to a central BS. This corresponds to a distributed BS scenario. The concept of distributed base stations (DBSs) and remote radio heads (RRHs) emerged to increase the coverage and capacity of wireless networks in a cost-effective way. It consists of a centrally located BS enclosure connected to RRHs via fiber optic cables [66]. In the existing literature, the terms distributed base station and distributed antenna system (DAS) are used interchangeably. DBSs were initially proposed to enhance indoor coverage of cellular systems where a building is treated as a single cell with several distributed antennas rather than either multiple pico cells each with a dedicated antenna or as a single cell with one central antenna [67]. The DBS approach allows avoiding excessive handovers in the first case and significant fading in the latter. The coverage and capacity of DBSs in an indoor WCDMA system was investigated in Ref. [68] for several types of antennas. In a multicell system with DBSs, it was shown that maximum ratio combining (MRC) in the uplink achieves a considerable capacity and coverage enhancement, but simultaneous transmission in the downlink reduces performance since it increases the intercell interference [69]. A solution for this problem was proposed in Ref. [70], where it was found that selecting only the RRH with best channel to the user ensures the best downlink performance with DBSs. A similar conclusion was reached in Ref. [71] where transmitting from the RRH with the best channel was shown to outperform the case of using the RRH as a relay while transmitting the signal directly from the BS. These results are validated from an information theoretic standpoint in Refs [72, 73], where selective transmission (from only the RRH with best channel to user) in the downlink was compared to maximum ratio transmission (using all the RRHs).

It should be noted that, in a practical scenario, installing RRHs at desired locations (e.g., equidistant along the cell boundary) might not be possible. Therefore, the performance of random placement of RRHs throughout the cell is investigated in Refs [74, 75], in terms of outage probability, as a lower bound on the actual performance. Interestingly, it is found that as the number of RRHs increases, the performance converges to that of regularly deployed RRHs. In fact, in the case of both fixed and random RRH locations, the gains achieved by a DBS system are shown to increase with the number of RRHs up to a certain limit where the gain obtained after using an additional RRH is negligible. This limit is considered to be four and seven RRHs per cell in Refs [70] and [72], respectively, for the regular RRH positions, and seven in Ref. [75] for the random RRH positions.

2.3 SCHEDULING IN MULTICELL SCENARIOS

In Sections 2.1 and 2.2, the main focus is on a single cell scenario and hence intercell interference is not considered. The emphasis is on mitigating intercell interference,

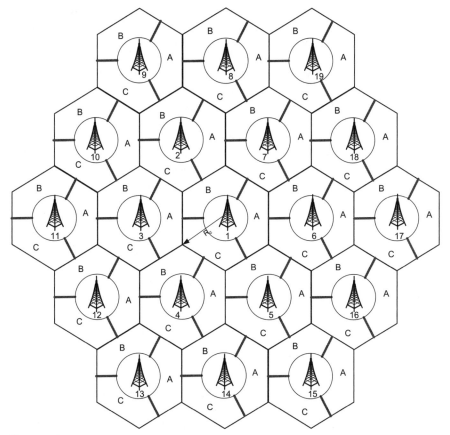

Figure 2.2   Fractional frequency reuse.

since intracell interference is not an issue in OFDMA due to the orthogonality of the subcarriers and the exclusivity of subcarrier allocations in each cell. To limit the effects of interference in multicell scenarios, several techniques for reusing the radio frequencies are investigated in the literature. Static reuse schemes are based on fractional frequency reuse (FFR) where a cell is divided into an inner area with the same frequencies reused in all cells and an outer area where a subset of the frequencies is reused [76]. Such an FFR scheme is illustrated in Fig. 2.2, where the entire bandwidth is divided into four segments. Part of the RBs is used with reuse of 1 in the cell center region, whereas reuse 3 is applied in regions A, B, and C [77]. An extension of the FFR reuse scheme for multicell scheduling in OFDMA networks is presented in Ref. [78], where each cell is subdivided into three regions: an inner region with the same frequencies reused in all cells (reuse 1), a middle region with an FFR with reuse 3, and an outer region with an FFR with reuse 9. Better interference mitigation is shown to be achieved with the enhanced FFR scheme of Ref. [78].

More efficient schemes consist of dynamic frequency reuse where all the frequencies are allowed to be used in all cells and elaborate techniques are applied for interference mitigation or avoidance. In Ref. [79], downlink scheduling in OFDMA is used with each BS randomly turning on or off certain subcarriers to mitigate interference. In Ref. [80], downlink transmit power allocation in a multicell wireless network under a sum-capacity maximization criterion and peak power constraints at each BS is investigated. It is shown that the optimal power control is binary (on–off) for two cells, and that binary power control yields a negligible capacity loss for more than two cells. However, scheduling is not used with power control in Ref. [80]. This has been explored in Ref. [81], where a distributed algorithm for power allocation and scheduling in multicell networks is proposed. Intercell coordination is applied to maximize the overall system capacity by deactivating cells that do not offer enough capacity to outweigh interference caused to the network. A single frequency is considered in Ref. [81], and the application of the approach to OFDMA networks necessitates the use of subcarrier allocation to exploit the frequency diversity gain. In Ref. [82], resource allocation is considered in the downlink of multicell OFDMA systems without BS cooperation. A price that increases with the transmit power is used in order to reduce the interference. The prices are used as a sort of power control scheme to reduce transmission power.

2.3.1 Multicell Scheduling in LTE

LTE performs interference mitigation using power control. The LTE power control scheme is detailed in Ref. [83], and its evaluation via simulations is described in Ref. [84]. LTE power control does not necessitate coordination between BSs for the purpose of interference mitigation. However, LTE allows communication between BSs over the X2 interface [85]. Intercell interference coordination (ICIC) between LTE BSs can be performed using the overload indicator (OI) and the high interference indicator (HII). The OI indicates the interference level received on each RB in the cell sending the OI, whereas the HII indicates the occurrence (or not) of high interference on each RB [85]. The BS receiving the OI and HII would then try to perform scheduling while avoiding allocation on the RBs subjected to high interference in its neighbor cells.

In the current LTE standard, the minimum latency for the exchange of information between BSs is 20 ms, whereas RBs are allocated on a 1 ms basis (duration of one TTI). This makes real-time processing of interference cancellation data from adjacent BSs unfeasible [86]. In Ref. [87], LTE resource allocation is performed in two levels: a fast (1 ms TTI) intracell level that does not involve ICIC, and a slow intercell level (duration of several TTIs) involving ICIC. The HII indicator is used in a proactive way in order to signal "protected bands" not to be allocated in neighbor cells. When the proactive scheme fails to select perfectly nonoverlapping protected bands, the OI indicator is used to request neighbor cells to refrain from scheduling UEs detected causing high interference. This scheme is shown in Ref. [87] to lead to enhanced performance. However, it assumes that all RBs are allocated to a single user per cell at a given TTI.

Other interference coordination and cancellation techniques applicable to the current LTE standard are surveyed in Ref. [86] (applicable to both uplink and downlink, unless otherwise specified): power control, static and adaptive fractional frequency reuse, MIMO (LTE downlink), multiuser MIMO (LTE uplink), space division multiple access (SDMA), interference cancellation (LTE uplink), opportunistic spectrum access, organized beamforming, sphere decoding (LTE uplink), and dirty paper decoding (LTE uplink). It should be noted that scheduling algorithms in general, including ICIC algorithms, are out of the scope of the LTE standard. The LTE standard has been developed to support a wide range of interference coordination approaches while allowing various types of scheduler operation, including channel state dependent (opportunistic) schedulers [87]. However, the benefits of channel aware scheduling discussed in Sections 2.1 and 2.2 tend to limit the additional benefits of ICIC [87], and the benefits of ICIC may not justify the added complexity of intercell communications [77].

In LTE, although the standard does not explicitly prescribe the timescale at which ICIC should operate, ICIC is currently performed on a scale of tens to hundreds of milliseconds, whereas fast scheduling is performed each millisecond [87]. Fast ICIC performed on the millisecond scale is beyond the current LTE standard and may be applicable to LTE-Advanced, where faster ICIC control is considered.

In Ref. [88], a beam coordination approach is proposed for LTE-Advanced, where each BS groups users served by the same beam in a way to reduce interference and enhance cell rate and edge users rate. The approach of Ref. [88] is based on feedback from users in each cell, and does not require inter-BS communication. It can be considered to be at the low end of coordinated multipoint (CoMP), which is a form of network MIMO. Major research is ongoing for network MIMO within the framework of LTE-Advanced, such as [89–92]:

- Synchronization of jointly processed terminals in time and frequency, and detection under synchronization offsets.

- Multisector channel estimation, feedback of CSI to BSs, and impact of imperfect CSI on network MIMO.

- Performance of network MIMO under a limited backhaul infrastructure between cooperating BSs.

- Cooperative scheduling for network MIMO.

It should be noted that in the CoMP and network MIMO literature [93, 94], sometimes a central control unit is assumed to coordinate the actions of BSs. This corresponds to a hierarchical level higher than the BSs in the network organization. In this case, the centralized approach corresponds to a scenario where the user equipment (UE) estimates the channel information from all the cooperating BSs and feeds it back to the central control unit, where scheduling operations are performed accordingly [93, 94]. The scenario where BSs take actions without the presence of a central control unit is referred to as "decentralized." In a decentralized scenario, the UE feeds back the channel information to all the cooperating BSs. Therefore, each BS gathers all the available feedback information, including those related to other BSs [93, 94].

2.4   SUMMARY

In this chapter, a survey of downlink resource allocation in OFDMA systems was presented. Three major topics were studied. The first is related to centralized scheduling within a single cell. Solutions based on instantaneous and ergodic scheduling were presented, in addition to a discussion of suboptimal scheduling techniques. The second topic is related to distributed scheduling where the subject of distributed antenna systems in OFDMA was analyzed. The third topic corresponds to multicell scheduling, where efficient interference mitigation techniques are needed in order to reduce the impact of intercell interference on the resource allocation process. Applications of the multicell scheduling schemes to LTE were presented, and the additional constraints imposed by the LTE standard were outlined.

CHAPTER 3

ERGODIC SUM-RATE MAXIMIZATION WITH CONTINUOUS RATES

In this chapter[1], the problem of uplink resource allocation in single cell OFDMA networks is discussed. In particular, the focus is on ergodic sum rate maximization and on deriving its optimal solution subject to per-user power and rate constraints. The presented optimal solution provides a bound on performance for practical resource allocation algorithms. In addition, the characteristics of the achievable rate region are determined. The chapter is structured as follows. Section 3.1 presents a review of the related literature and outlines the main topics of this chapter. The problem is formulated in Section 3.2 and the solution is presented in Section 3.3. The achievable rate region is determined in Section 3.4. Numerical results are presented and analyzed in Section 3.5. Finally, Section 3.6 summarizes the chapter.

3.1 BACKGROUND

Resource allocation in OFDMA has been widely investigated in the downlink both in the case of instantaneous rate maximization (e.g., [18–20, 23, 24]) and ergodic rate maximization (e.g., [26, 27]). The solution is generally divided into two parts: subcarrier allocation and power allocation. The solution described in Ref. [19] consists of allocating each subcarrier to the user with the best channel condition on that subcarrier and allocating power by water-filling over the subcarriers.

The scheduling problem in the uplink is more challenging than that in the downlink due to the distributed power constraint: in the downlink, the power has a centralized nature because power allocation is done at a central entity, the base station (BS), whereas in the uplink, the power has a distributive nature and should be

---

[1] This chapter is adapted, with permission from IET, from E. Yaacoub, A. M. El-Hajj, and Z. Dawy, Weighted Ergodic Sum-Rate Maximization in Uplink OFDMA and its Achievable Rate Region, *IET Communications*, 4(18), 2217–2229, 2010.

Resource Allocation in Uplink OFDMA Wireless Systems: Optimal Solutions and Practical Implementations, Elias E. Yaacoub and Zaher Dawy.

considered on a per-user basis. Instantaneous scheduling in the uplink was investigated in Refs [47, 95–97]. The allocation problem is divided into two subproblems in Ref. [95]. A greedy algorithm is proposed with water-filling used to allocate power for each user on its allocated subcarriers. Then using the marginal functions, an optimal (user, subcarrier) pair is found. Steps are repeated until all subcarriers are allocated. In Ref. [96], fairness is added to the approach of Ref. [95] by allocating subcarriers to a given user until its required rate is reached then the user is excluded from the allocation of the remaining subcarriers. The algorithm proposed in Ref. [47] has similar steps to that of Ref. [95], but differs in that it performs water-filling for each user on all unallocated subcarriers in addition to the subcarriers allocated to that user before searching for the optimal (user, subcarrier) pair. The algorithms in Refs [47, 95, 96] are suboptimal. In Ref. [97], instantaneous sum-rate maximization is formulated into a convex optimization problem and solved using a dual decomposition approach, then a set of suboptimal algorithms are presented and compared, due to the prohibitive complexity of implementing the optimal solution.

In this chapter, maximization of weighted uplink ergodic sum-rate in OFDMA systems is considered. In weighted ergodic sum-rate maximization, advantage is taken of the time dimension in the optimization in addition to the frequency and multiuser diversity dimensions [26]. Ergodic maximization was considered in Ref. [33] in the context of an ad hoc cognitive radio network. The total weighted rate in the system was considered with rate constraints on the primary users.

The main topics discussed in this chapter are summarized as follows:

1. Formulating and solving the ergodic weighted sum-rate maximization problem subject to per-user power and rate constraints in the uplink of OFDMA systems. Instantaneous sum-rate maximization is considered in Ref. [97], but the additional per-user rate constraints are not considered. Furthermore, using the time dimension in the ergodic optimization in addition to the frequency and multiuser diversity dimensions leads to an optimal solution not having a prohibitive implementation complexity. In Ref. [33], weighted ergodic sum-rate maximization in cognitive OFDMA radio networks is studied, with rate constraints protecting only the primary users. However, users are assumed to know the channel realizations on the links between all pairs of users over all subcarriers in order to compute the Lagrangian parameters and perform resource allocation in a distributed way. This necessitates a large overhead for exchanging the information between users and thus limits the practical implementation of the approach. In this chapter, a centralized OFDMA cellular system is considered where the BS is the sole entity responsible of resource allocation, and hence the channel estimation, the computation of the parameters, and the application of the algorithm are done at a single location. The estimation of the channel state information (CSI) between each user and the BS does not impose a practical limitation, since it is implemented in dynamic resource allocation for most state-of-the-art wireless communication systems. For example, in LTE, the BS extracts the necessary uplink CSI from the sounding reference signal (SRS) transmitted by the mobile in order to perform channel dependent scheduling [98].

2. Analyzing the duality gap of the proposed solution and showing that a zero duality gap is reached when the optimal dual solution is derived.

3. Determining the characteristics of the achievable rate region for an arbitrary number of users and investigating the effect of the per-user rate constraints on the achievable region. Numerical results illustrating the theoretical derivations are considered for the two-users rate region. Although OFDMA downlink rate regions are treated in Refs [24, 26, 99], per-user rate constraints are not considered and ergodicity is investigated only in Ref. [26] for the two-users case. In the OFDMA uplink, rate regions were traced out in Ref. [100] for the two-users case using a suboptimal algorithm. However, the approach of Ref. [100] does not investigate ergodic sum-rate maximization and does not consider per-user rate constraints in order to categorize their impact on the achievable rate region.

4. Discussing the practical implementation of the proposed solution by presenting online and offline techniques to compute the Lagrangian parameters and analyzing their convergence time. A relatively similar analysis is presented in Ref. [33] for cognitive OFDMA radio networks.

3.2 PROBLEM FORMULATION

A single cell uplink OFDMA system is considered, with K users and N subcarriers to be allocated. For each user k and subcarrier i, the channel gain and total noise power are respectively, denoted as $H_{k,i}$ and $\sigma_{k,i}^2$. The channel gain-to-noise ratio (CNR) is given by

$$g_{k,i} = \frac{H_{k,i}}{\sigma_{k,i}^2}, \quad k = 1, \ldots K; i = 1, \ldots, N \tag{3.1}$$

Without loss of generality, normalized values of $g_{k,i}$ are considered (i.e., $\sigma_{k,i}^2 = 1$ for all k and i). Let $\alpha_{k,i}$ be the binary decision variable of subcarrier allocation. If subcarrier i is allocated to user k then $\alpha_{k,i}$ is equal to 1, otherwise it is equal to 0. Since each subcarrier is exclusively allocated to one user, then

$$\sum_{k=1}^{K} \alpha_{k,i} \leq 1, \quad i = 1, \ldots, N \tag{3.2}$$

It should be noted that in this chapter, the term "rate region" is used instead of "capacity region", since the subcarriers are allocated exclusively to each user. The capacity regions for multiple access and broadcast channels have been investigated in Refs [37, 101], respectively. It was proven that the capacity is achieved when the same frequency range is shared with overlap by multiple users and successive decoding is applied. It was however shown in Ref. [102] that there is only a small range of frequency with overlapping when optimal power allocation is used. Thus, it was asserted in Ref. [19], based on the results of Refs [18, 103, 104], that optimal power allocation with exclusive dynamic subcarrier assignment can achieve a data transmission rate close to the channel capacity boundary. Therefore, the main interest in this book is in exclusive subcarrier assignment, since it is used in state-of-the-art wireless communication systems, for example, LTE and WiMAX, and it achieves results close to capacity while avoiding the complexity of successive multiuser decoding. Let $P_{k,i}$ be the power allocated to subcarrier i by user k. Then,

$$\sum_{i=1}^{N} P_{k,i} \leq P_{k,\max}, \quad k = 1, \ldots, K \tag{3.3}$$

This means that the power spent by the user over all its allocated subcarriers should be lower than its maximum transmission power $P_{k,\max}$. The total rate of user k is defined as follows:

$$R_k = \sum_{i=1}^{N} \alpha_{k,i} \log_2(1 + P_{k,i} g_{k,i}) \tag{3.4}$$

Consequently, the total system rate is given by

$$R(\mathbf{A}, \mathbf{P}) = \sum_{k=1}^{K} \sum_{i=1}^{N} \alpha_{k,i} \log_2(1 + P_{k,i} g_{k,i}) \tag{3.5}$$

where $\mathbf{A}$ is a $K \times N$ matrix of channel allocation indices $\alpha_{k,i}$, and $\mathbf{P}$ is a $K \times N$ matrix of allocated powers $P_{k,i}$.

The problem of the maximization of the weighted ergodic sum-rate can be formulated as follows:

$$\max_{\mathbf{A},\mathbf{P}} \mathbb{E}_g \left\{ \sum_{k=1}^{K} \pi_k \sum_{i=1}^{N} \alpha_{k,i} \log_2(1 + P_{k,i} g_{k,i}) \right\}$$

subject to

$$\mathbb{E}_g \left\{ \sum_{i=1}^{N} P_{k,i} \right\} \leq P_{k,\max}, \quad \text{for all } k \tag{3.6}$$

$$\mathbb{E}_g \left\{ \sum_{i=1}^{N} \alpha_{k,i} \log_2(1 + P_{k,i} g_{k,i}) \right\} \geq R_{k,\min}, \quad \text{for all } k$$

where $\mathbb{E}\{.\}$ is the expectation operator, π_k is the weight given to the rate of user k, $P_{k,\max}$ and $R_{k,\min}$ are the maximum transmission power and the minimum target rate of user k, respectively. Thus, $\mathbb{E}_g\{.\}$ denotes the expectation over the realizations of the channel gain probability distribution function (pdf).

The weights are chosen such that $\sum_{k=1}^{K} \pi_k = 1$. In an OFDMA-based wireless communication system, these weights are generally handed down from the MAC layer to the PHY layer scheduling routine on a per-frame (or longer) basis [26]. They are used to give the rates of certain users more importance in the maximization of (3.6) thus providing a notion of fairness.

The problem in (3.6) is nonconvex due to the discrete set of values taken by $\alpha_{k,i}$. Thus, the condition on the $\alpha_{k,i}$s is relaxed by allowing them to take any value in the interval $[0, 1]$ instead of the set $\{0, 1\}$, which is equivalent to allowing time-sharing of a single subcarrier between different users. By "time-sharing" we mean that several users can transmit on a given subcarrier during a given scheduling interval, with each user transmitting alone for a fraction of the interval. This corresponds to a sort of time division multiple access (TDMA) subdivision of the scheduling time unit and is not to be confused with overlapping transmission. Letting $f_{k,i} = \alpha_{k,i} P_{k,i}$, the problem in (3.6) can be reformulated as follows:

$$\max_{\mathbf{A},\mathbf{F}} \mathbb{E}_g\left\{ \sum_{k=1}^{K} \pi_k \sum_{i=1}^{N} \alpha_{k,i} \log_2\left(1 + \frac{f_{k,i}}{\alpha_{k,i}} g_{k,i} \right) \right\}$$

subject to

$$\mathbb{E}_g\left\{ \sum_{i=1}^{N} f_{k,i} \right\} \le P_{k,\max}, \quad \text{for all } k \tag{3.7}$$

$$\mathbb{E}_g\left\{ \sum_{i=1}^{N} \alpha_{k,i} \log_2\left(1 + \frac{f_{k,i}}{\alpha_{k,i}} g_{k,i} \right) \right\} \ge R_{k,\min}, \quad \text{for all } k$$

The problem in (3.7) is convex with a concave objective function since expectation preserves convexity and a function of the form $a \log_2(1 + b/a)$ is known to be concave [34]. Note that in the limit, it can be written that $\lim_{a \to 0} a \log_2(1 + b/a) = 0$. The problem in (3.7) is equivalent to the original problem in (3.6) when the condition on $\alpha_{k,i}$s is relaxed: for each user k and subcarrier i, finding the optimal pair $(\alpha_{k,i}^*, f_{k,i}^*) = (\alpha_{k,i}^*, \alpha_{k,i}^* P_{k,i}^*)$ leads to the same solution as finding $(\alpha_{k,i}^*, P_{k,i}^*)$.

3.3 PROBLEM SOLUTION

In some situations, finding the solution of a convex optimization problem might not be straightforward. In this case, a possible approach would be to formulate the dual of the problem (the initial problem is called the "primal"). In general, the dual incorporates the constraints into the objective function to optimize in the primal problem, by introducing the Lagrangian parameters. If the dual problem can be solved, and when the optimal dual solution is reached, it may or may not correspond to the optimal primal solution. The "gap" between the primal and dual solutions is called the "duality gap." When this gap is zero, this means that the dual solution corresponds exactly to the primal solution and hence the primal problem is solved. Sometimes, the gap may be small but nonzero. In this case, the dual solution corresponds to a "good", but not "optimal", primal solution. In this section, the weighted ergodic sum-rate maximization primal problem is solved by defining the dual problem using Lagrangian parameters, solving the dual problem, then showing that the duality gap is zero. The interested reader may find more details about convex optimization and dual formulations, for example, in Refs [34, 35] (Chapters 4 and 5 in Ref. [34] contain the most relevant background to Chapters 3–5 of this book.)

3.3.1   Solution of the Dual Problem

In order to reach the solution of the problem defined in (3.7), the Lagrangian is defined. It entails two vectors of Lagrange multipliers λ and μ corresponding to the power and rate constraints, respectively.

$$
\begin{aligned}
L(\mathbf{A}, \mathbf{F}, \lambda, \mu) = \mathbb{E}_g & \left\{ \sum_{k=1}^{K} \pi_k \sum_{i=1}^{N} \alpha_{k,i} \log_2 \left(1 + \frac{f_{k,i}}{\alpha_{k,i}} g_{k,i} \right) \right\} \\
& - \sum_{k=1}^{K} \lambda_k \left(\mathbb{E}_g \left\{ \sum_{i=1}^{N} f_{k,i} \right\} - P_{k,\max} \right) \\
& - \sum_{k=1}^{K} \mu_k \left(R_{k,\min} - \mathbb{E}_g \left\{ \sum_{i=1}^{N} \alpha_{k,i} \log_2 \left(1 + \frac{f_{k,i}}{\alpha_{k,i}} g_{k,i} \right) \right\} \right)
\end{aligned}
\tag{3.8}
$$

The Lagrangian dual function is then given by

$$
D(\lambda, \mu) = \max_{\mathbf{A}, \mathbf{F}} L(\mathbf{A}, \mathbf{F}, \lambda, \mu)
\tag{3.9}
$$

The optimization dual problem is given by

$$
\min_{\lambda \geq 0, \mu \geq 0} D(\lambda, \mu)
\tag{3.10}
$$

The Lagrangian dual can be rewritten, after interchanging the expectation and the summation, as follows:

$$
\begin{aligned}
D(\lambda, \mu) = \max_{\mathbf{A}, \mathbf{F}} & \sum_{k=1}^{K} \sum_{i=1}^{N} \mathbb{E}_g \left\{ (\pi_k + \mu_k) \alpha_{k,i} \log_2 \left(1 + \frac{f_{k,i}}{\alpha_{k,i}} g_{k,i} \right) - \lambda_k f_{k,i} \right\} \\
& + \sum_{k=1}^{K} \lambda_k P_{k,\max} - \sum_{k=1}^{K} \mu_k R_{k,\min} \\
= \max_{\mathbf{A}} & \sum_{k=1}^{K} \sum_{i=1}^{N} \max_{\mathbf{P}} \mathbb{E}_g \left\{ \alpha_{k,i} \left((\pi_k + \mu_k) \log_2 (1 + P_{k,i} g_{k,i}) - \lambda_k P_{k,i} \right) \right\} \\
& + \sum_{k=1}^{K} \lambda_k P_{k,\max} - \sum_{k=1}^{K} \mu_k R_{k,\min} \\
= \max_{\mathbf{A}} & \sum_{k=1}^{K} \sum_{i=1}^{N} \max_{\mathbf{P}} \mathbb{E}_g \left\{ \alpha_{k,i} \Phi(P_{k,i}) \right\} + \sum_{k=1}^{K} \lambda_k P_{k,\max} - \sum_{k=1}^{K} \mu_k R_{k,\min}
\end{aligned}
\tag{3.11}
$$

with $\Phi(P_{k,i})$ defined as follows:

$$
\Phi(P_{k,i}) = (\pi_k + \mu_k) \log_2 (1 + P_{k,i} g_{k,i}) - \lambda_k P_{k,i}
\tag{3.12}
$$

To maximize (3.11) for any given $\alpha_{k,i}$, (3.12) is differentiated with respect to $P_{k,i}$ and the result is set to 0. This yields

$$P_{k,i}^* = \left[\frac{\pi_k + \mu_k}{\ln(2)\lambda_k} - \frac{1}{g_{k,i}} \right]^+ \tag{3.13}$$

where $[y]^+ = \max(0, y)$.

The solution in (3.13) corresponds to the maximum per each fading state realization and hence for the expectation over g [33]. This solution corresponds to the standard water-filling solution and depends on the multiplier λ_k associated with the per user power constraint, the multiplier μ_k associated with the per user rate constraint, and the channel gain of user k on subcarrier i.

The second maximization can now be treated by replacing, in (3.11), $P_{k,i}$ with the optimal value. This yields

$$D(\lambda, \mu) = \max_{\mathbf{A}} \sum_{k=1}^{K} \sum_{i=1}^{N} \mathbb{E}_g \left\{ \alpha_{k,i} \Phi(P_{k,i}^*) \right\} + \sum_{k=1}^{K} \lambda_k P_{k,\max} - \sum_{k=1}^{K} \mu_k R_{k,\min} \tag{3.14}$$

The variable $\alpha_{k,i}$ is also a function of each realization of g. Furthermore, by definition $\sum_{k=1}^{K} \alpha_{k,i} \leq 1$. Hence, the following can be noted:

$$\sum_{k=1}^{K} \alpha_{k,i} \Phi(P_{k,i}^*) \leq \max_k \Phi(P_{k,i}^*), \quad \text{for all } i \tag{3.15}$$

Although time-sharing is allowed, (3.15) corresponds to allocating each subcarrier i to the user that maximizes $\Phi(P_{k,i}^*)$. This corresponds to a "winner takes all" on each subcarrier and the maximization over $\mathbf{A}$ reduces to the following rule

$$\alpha_{k,i}^* = \begin{cases} 1, & \text{if } k = k_i^* \text{ for } i = 1, ..., N \\ 0, & \text{otherwise} \end{cases} \tag{3.16}$$

where

$$k_i^* = \arg \max_k \Phi(P_{k,i}^*) \tag{3.17}$$

Hence, although the problem was relaxed by allowing time-sharing, the optimal solution reached consists of exclusive subcarrier allocation. As a result,

$$f_{k,i}^* = P_{k,i}^* \alpha_{k,i}^* = \begin{cases} P_{k,i}^*, & \text{if } \alpha_{k,i}^* = 1 \\ 0, & \text{otherwise} \end{cases} \tag{3.18}$$

Finally, to complete the solution, the vector of geometric multipliers associated with the power and rate constraints needs to be determined. Determining the multipliers can be done using an iterative subgradient method with each iteration on the multipliers λ_k and μ_k given by

$$
\begin{aligned}
\lambda_k^{n+1} &= \left[\lambda_k^n - \delta_n G_{\lambda_k}^n\right]^+ \\
\mu_k^{n+1} &= \left[\mu_k^n - \delta_n G_{\mu_k}^n\right]^+
\end{aligned}
\tag{3.19}
$$

where the superscript n denotes the index of the iteration and G_λ^n and G_μ^n denote the subgradients, which are taken as

$$
\begin{aligned}
G_{\lambda_k}^n &= P_{k,\max} - \mathbb{E}_g\left\{\sum_{i=1}^N f_{k,i}^*\right\} \\
G_{\mu_k}^n &= \mathbb{E}_g\left\{\sum_{i=1}^N \alpha_{k,i}^* \log_2\left(1 + \frac{f_{k,i}^*}{\alpha_{k,i}^*} g_{k,i}\right)\right\} - R_{k,\min}
\end{aligned}
\tag{3.20}
$$

The step size δ_n is taken of the form $\delta_n = a/\sqrt{n}$, where a is a positive constant. This chosen step size is guaranteed to lead to convergence since it obeys the nonsummable diminishing rule [35]:

$$
\lim_{n \to \infty} \delta_n = 0, \qquad \sum_{n=1}^{\infty} \delta_n = \infty.
\tag{3.21}
$$

3.3.2   Duality Gap Analysis

Using the obtained power and subcarrier allocation, the dual solution can be formulated as follows:

$$
\begin{aligned}
d^* &= \sum_{k=1}^K \sum_{i=1}^N \mathbb{E}_g\left\{(\pi_k + \mu_k^*)\alpha_{k,i}^* \log_2\left(1 + \frac{f_{k,i}^*}{\alpha_{k,i}^*} g_{k,i}\right) - \lambda_k^* f_{k,i}^*\right\} \\
&\quad + \sum_{k=1}^K \lambda_k^* P_{k,\max} - \sum_{k=1}^K \mu_k^* R_{k,\min}
\end{aligned}
\tag{3.22}
$$

The power constraint states that the average of the total power used by each user at each scheduling interval should be less than the user's target maximum power $P_{k,\max}$. The average power of user k is denoted by

$$
P_k^{\mathrm{avg}}(\lambda_k^*, \mu_k^*) = \mathbb{E}_g\left\{\sum_{i=1}^N f_{k,i}^*\right\}
\tag{3.23}
$$

The minimum rate constraint states that the average rate of each user should be greater than the user's target minimum rate $R_{k,\min}$. The average rate of user k is denoted by

$$
R_k^{\mathrm{avg}}(\lambda_k^*, \mu_k^*) = \mathbb{E}_g\left\{\sum_{i=1}^N \alpha_{k,i}^* \log_2\left(1 + \frac{f_{k,i}^*}{\alpha_{k,i}^*} g_{k,i}\right)\right\}
\tag{3.24}
$$

The dual solution can therefore be rewritten as

$$d^* = \sum_{k=1}^{K} \sum_{i=1}^{N} \mathbb{E}_g \left\{ \pi_k \alpha_{k,i}^* \log_2 \left(1 + \frac{f_{k,i}^*}{\alpha_{k,i}^*} g_{k,i} \right) \right\}$$

$$+ \sum_{k=1}^{K} \lambda_k^* \left(P_{k,\max} - P_k^{\mathrm{avg}}(\lambda_k^*, \mu_k^*) \right) + \sum_{k=1}^{K} \mu_k^* \left(R_k^{\mathrm{avg}}(\lambda_k^*, \mu_k^*) - R_{k,\min} \right)$$

$$\tag{3.25}$$

The primal solution can be written as

$$p^* = \mathbb{E}_g \left\{ \sum_{k=1}^{K} \sum_{i=1}^{N} \pi_k \alpha_{k,i}^* \log_2 \left(1 + \frac{f_{k,i}^*}{\alpha_{k,i}^*} g_{k,i} \right) \right\} \tag{3.26}$$

From the weak duality theorem [35], given a problem with a dual solution d^* and a primal solution p^*, then

$$p^* \le d^* \tag{3.27}$$

The duality gap is given by

$$d^* - p^* =$$

$$\sum_{k=1}^{K} \sum_{i=1}^{N} \pi_k \mathbb{E}_g \left\{ \alpha_{k,i}^* \log_2 \left(1 + \frac{f_{k,i}^*}{\alpha_{k,i}^*} g_{k,i} \right) \right\} + \sum_{k=1}^{K} \lambda_k^* \left(P_{k,\max} - P_k^{\mathrm{avg}}(\lambda_k^*, \mu_k^*) \right)$$

$$- \mathbb{E}_g \left\{ \sum_{k=1}^{K} \sum_{i=1}^{N} \pi_k \alpha_{k,i}^* \log_2 \left(1 + \frac{f_{k,i}^*}{\alpha_{k,i}^*} g_{k,i} \right) \right\} + \sum_{k=1}^{K} \mu_k^* \left(R_k^{\mathrm{avg}}(\lambda_k^*, \mu_k^*) - R_{k,\min} \right)$$

$$\tag{3.28}$$

From this equation, it can be easily shown that the duality gap is zero whenever

$$\lambda_k^* \left(P_{k,\max} - P_k^{\mathrm{avg}}(\lambda_k^*, \mu_k^*) \right) = 0, \quad \text{for all } k \tag{3.29}$$

and

$$\mu_k^* \left(R_k^{\mathrm{avg}}(\lambda_k^*, \mu_k^*) - R_{k,\min} \right) = 0, \quad \text{for all } k \tag{3.30}$$

To set (3.29) to zero, the condition $P_k^{\mathrm{avg}} = P_{k,\max}$ must be satisfied for all k, since having $\lambda_k^* = 0$ will lead to an infinite power in (3.13). As an exception, it might occur that $\lambda_k^* = 0$ only in the case where the optimal solution mandates that user k is not allocated any subcarriers and hence its power is zero according to (3.15). This situation can occur when $\pi_k = 0$ for example. On the other hand, to satisfy (3.30), the condition $R_k^{\mathrm{avg}} = R_{k,\min}$ must be met only for users that have $R_k^{\mathrm{avg}} < R_{k,\min}$ in the optimal solution of the weighted ergodic sum-rate maximization problem without rate constraints, that is, when the rate constraints are not explicitly imposed. Users achieving $R_k^{\mathrm{avg}} > R_{k,\min}$ must have $\mu_k^* = 0$ in order to reach a zero duality gap, since forcing these users to achieve only the minimum rate leads to a suboptimal solution, due to reducing their contribution in the weighted ergodic sum-rate.

3.3.3 Complexity Analysis

The proposed solution requires a complexity of $\mathcal{O}(INK)$ in the Lagrangian parameters computation phase, with I the number of iterations. Afterwards, the complexity becomes $\mathcal{O}(NK)$. The optimal ergodic sum-rate maximization solution is simpler to implement than the optimal instantaneous sum-rate maximization, which corresponds to (3.7) without the expectation. In fact, instantaneous sum-rate maximization requires a complexity of $\mathcal{O}(INK)$ at each scheduling interval, since the Lagrangian parameters need to be computed at every scheduling instant in order to obtain the optimal performance. More details about this topic are given in Section 3.5.2.

3.3.4 Solution Approach in a MIMO Scenario

The derivations obtained in this chapter were based on OFDMA systems with single transmit and receive antennas. Multiple input multiple output (MIMO) systems are becoming common in state-of-the art wireless communication systems, and will be an integral part of fourth generation (4G) systems [93, 105–107]. Hence, extending the results of this chapter to systems with multiple transmit and receive antennas, that is, MIMO–OFDMA, is an interesting problem for further investigation. In this section, an overview of the solution approach to be followed in the case of MIMO is presented.

In MIMO–OFDMA, let N_r denote the number of receive antennas (at the BS) and by N_t the number of transmit antennas (at the mobile device). In this case, the frequency-selective channel can be modeled by a $N_r \times N_t$ channel matrix per subcarrier. Thus, using singular value decomposition, up to $\min(N_r, N_t)$ spatial gains are obtained per subcarrier. Hence, instead of the K CNR values per subcarrier in the scheduling problem of Section 3, $K \min(N_r, N_t)$ values should be included in the scheduling problem. Then, power allocation using multilevel water-filling should be applied across all spatial gains of all subcarriers, that is, the application of (3.13) is performed on $K \min(N_r, N_t)$ channel gains for each subcarrier. Afterwards, the subcarrier should be assigned to the user that maximizes the marginal dual, which in this case corresponds to the MIMO capacity per-subcarrier. In other words, the power allocation obtained is used in (3.12), where the MIMO capacity is computed by replacing the capacity over a single channel per subcarrier by the total capacity over the $K \min(N_r, N_t)$ spatial channels corresponding to that subcarrier. Then the subcarrier is allocated to the user maximizing (3.15).

The extension related to MIMO is mainly applicable in a centralized scheduling scenario where the BS tightly controls the resource allocation process, since it requires a large amount of channel state information for all users on all subcarriers and between all antenna pairs. Although deriving the optimal solution has a complexity increasing with the number of antennas, such a solution presents a benchmark to which practical suboptimal scheduling algorithms can be compared.

3.4 ACHIEVABLE RATE REGION

In this section, the objective is to determine the characteristics of the K-user achievable rate region for ergodic weighted sum-rate maximization without rate constraints.

Then the impact of the per user rate constraints in reducing the size of this region is investigated. Finally, the results are applied to the special case of the two-user rate region.

3.4.1   K-user Achievable Rate Region without Rate Constraints

Lemma 3.1.   *The time division multiple access solutions for all possible weight vectors $\{\pi_k\}_{k=1}^{K}$ are along a hyperplane in $\mathcal{R}^{+K}$, where $\mathcal{R}$ is the set of real numbers.*

Proof. Let the point A_{k^*} correspond to the TDMA solution for $\pi_{k^*} = 1$ and $\pi_k = 0$ for all $k \neq k^*$, that is, A_{k^*} corresponds to allocating all the subcarriers to user k^* all the time. Clearly, A_{k^*} lies on the axis R_{k^*} (corresponding to the rate of user k^*) in $\mathcal{R}^{+K}$. The TDMA solution for a weight vector $\{\pi_k\}_{k=1}^{K}$ consists of allocating all the subcarriers to user k for a fraction π_k of the time. The point A corresponding to this solution is a linear combination $OA = \sum_{k=1}^{K} \pi_k OA_k$ and thus lies on the hyperplane containing the K points $OA_k, k = 1, \cdots, K$.   □

Theorem 3.1.   *Optimal subcarrier and power allocation achieves better performance than TDMA with optimal power allocation.*

Proof. It is clear that

$$
\begin{aligned}
\max_{\mathbf{A},\mathbf{P}} \mathbb{E}_g &\left\{ \sum_{k=1}^{K} \pi_k \sum_{i=1}^{N} \alpha_{k,i} \log_2(1 + P_{k,i} g_{k,i}) \right\} \\
&\geq \max_{\mathbf{P}} \mathbb{E}_g \left\{ \sum_{k=1}^{K} \pi_k \sum_{i=1}^{N} \log_2(1 + P_{k,i} g_{k,i}) \right\}
\end{aligned}
\tag{3.31}
$$

where the right-hand side of (3.31) corresponds to allocating all the subcarriers exclusively to user k for a fraction π_k of the time, and the left-hand side corresponds to the solution derived in Section 3.3.   □

Theorem 3.2.   *The achievable rate region of the weighted ergodic sum-rate maximization is convex.*

Proof. Denoting by S the set of achievable rate vectors for users 1, ..., K, then for any $X, Y \in S^2$ and $\beta \in [0, 1], Z = \beta X + (1 - \beta)Y$ is achievable by time-sharing, and hence $Z \in S$. Consequently S is a convex set.   □

Lemma 3.2.   *Without rate constraints, the set of achievable rate vectors $\{R_k^{\text{avg}}\}_{k=1}^{K}$ for a given weight vector $\{\pi_k\}_{k=1}^{K}$ is along a line passing through the origin, the TDMA solution, and the optimal solution.*

Proof. The origin O is obviously a feasible (trivial) solution. The optimal solution D is the best feasible solution. Hence, any point on OD is achievable by time-sharing. Consequently, by Theorems 3.1 and 3.2, the line segment OD intersects the TDMA hyperplane on a single point C. Thus, C is a feasible solution for the weight vector $\{\pi_k\}_{k=1}^K$. Let C' be the TDMA solution for a given $\{\pi_k\}_{k=1}^K$. It corresponds to allocating all the subcarriers to user k for a fraction π_k of the time. Since C is also a solution for $\{\pi_k\}_{k=1}^K$ and C is on the TDMA hyperplane, then $C = C'$ and hence the trivial solution O, the TDMA solution C, and the optimal solution D are aligned.   □

Lemma 3.3.   *The point D corresponding to the optimal solution $\{R_k^{*avg}\}_{k=1}^K$ for a given weight vector $\{\pi_k\}_{k=1}^K$ is the intersection of the line corresponding to $\{\pi_k\}_{k=1}^K$ (passing through the origin and the TDMA solution) with the boundary of the convex set of feasible solutions $\mathcal{S}$.*

Proof. Straightforward from Lemma 3.2.   □

3.4.2   K-user Achievable Rate Region with Rate Constraints

The following two corollaries describe the effect of the rate constraints on the achievable rate region. Their proofs are trivial and hence are omitted.

Corollary 3.1.   *A rate constraint for a user k is represented by a hyperplane defined by $R_k = R_{k,min}$ which is normal to the rate axis R_k corresponding to user k in $\mathcal{R}^{+K}$.*

Corollary 3.2.   *The achievable rate region with a given set of rate constraints $\mathcal{S}_c$ is the intersection of the achievable rate region without rate constraints $\mathcal{S}$ with the polyhedron defined by the intersection of the halfspaces $R_k \geq R_{k,min}$ bounded by the hyperplanes $R_k = R_{k,min}$.*

The distance presented in Definition 3.1 below is required in the proof of Theorem 3.3 presented next.

Definition 3.1.   *Let $D_\pi(X, Y)$ be defined, for a given $\{\pi_k\}_{k=1}^K$, as*

$$D_\pi(X, Y) = \left| \sum_{k=1}^K \pi_k R_{k,X} - \sum_{k=1}^K \pi_k R_{k,Y} \right| \tag{3.32}$$

*such that the vectors **OX** and **OY** in $\mathcal{R}^K$ are defined by $\mathbf{OX} = \{R_{k,X}\}_{k=1}^K$ and $\mathbf{OY} = \{R_{k,Y}\}_{k=1}^K$, and $|\cdot|$ is the absolute value operation.*

Corollary 3.3.   *D_π is a distance over $\mathcal{R}^K$.*

Proof. To prove Corollary 3.3, we show that D_π satisfies all the properties of a distance over $\mathcal{R}^{+K}$. In fact,

$$D_\pi(X, Y) \geq 0$$
$$D_\pi(X, X) = 0$$
$$D_\pi(X, Y) = D_\pi(Y, X)$$

$$
\begin{aligned}
D_\pi(X, Z) &= \left| \sum_{k=1}^{K} \pi_k R_{k,X} - \sum_{k=1}^{K} \pi_k R_{k,Z} \right| \\
&= \left| \sum_{k=1}^{K} \pi_k R_{k,X} - \sum_{k=1}^{K} \pi_k R_{k,Y} + \sum_{k=1}^{K} \pi_k R_{k,Y} - \sum_{k=1}^{K} \pi_k R_{k,Z} \right| \\
&\leq \left| \sum_{k=1}^{K} \pi_k R_{k,X} - \sum_{k=1}^{K} \pi_k R_{k,Y} \right| + \left| \sum_{k=1}^{K} \pi_k R_{k,Y} - \sum_{k=1}^{K} \pi_k R_{k,Z} \right| \\
&= D_\pi(X, Y) + D_\pi(Y, Z)
\end{aligned}
\tag{3.33}
$$

which completes the proof.   □

The next two theorems (Theorems 3.3 and 3.4) are essential in determining the impact of the per user rate constraints on the achievable rate region.

Theorem 3.3. *The optimal solution in the presence of rate constraints is represented by the point in the set $\mathcal{S}_c$, defined in Corollary 3.2, that is nearest in terms of the distance D_π, to the point representing the optimal solution without rate constraints in the set $\mathcal{S}$.*

Proof. Given $\{\pi_k\}_{k=1}^{K}$, let point X be the optimal solution in $\mathcal{S}$. Furthermore, given a set of minimum rate constraints $\{R_k^{\text{avg}} \geq R_{k,\min}\}_{k=1}^{K}$, let point Z be the optimal solution in $\mathcal{S}_c$ and let $Y \in \mathcal{S}_c$ be such that $D_\pi(X, Y) = \min_{Y' \in \mathcal{S}_c} D_\pi(X, Y')$. Theorem 3.3 is now proven by contradiction: since Z corresponds to the point in $\mathcal{S}_c$ maximizing the weighted sum-rate, then $\sum_{k=1}^{K} \pi_k R_{k,Z} \geq \sum_{k=1}^{K} \pi_k R_{k,Y}$. But since Y is the nearest point to X in $\mathcal{S}_c$, then

$$D_\pi(X, Z) \geq^{(a)} D_\pi(X, Y)$$

$$
\left| \sum_{k=1}^{K} \pi_k R_{k,X} - \sum_{k=1}^{K} \pi_k R_{k,Z} \right| \geq^{(b)} \left| \sum_{k=1}^{K} \pi_k R_{k,X} - \sum_{k=1}^{K} \pi_k R_{k,Y} \right|
$$

$$
\sum_{k=1}^{K} \pi_k R_{k,X} - \sum_{k=1}^{K} \pi_k R_{k,Z} \geq^{(c)} \sum_{k=1}^{K} \pi_k R_{k,X} - \sum_{k=1}^{K} \pi_k R_{k,Y}
\tag{3.34}
$$

$$
\sum_{k=1}^{K} \pi_k R_{k,Z} \leq^{(d)} \sum_{k=1}^{K} \pi_k R_{k,Y}
$$

where (a) is a result of the assumption that Y is nearest to X, (b) is obtained using the definition of D_π, (c) is reached since X is the solution maximizing the weighted sum-rate without constraints, and (d) is a direct consequence of (c). Clearly, the final result in (d) contradicts the assumption that Z is the optimal solution in $\mathcal{S}_c$. Hence, Y is the optimal solution, which completes the proof. ☐

Theorem 3.4.   *The point corresponding to the solution maximizing the weighted sum-rate with rate constraints is on the boundary of the convex region $\mathcal{S}$ corresponding to optimal solutions without rate constraints.*

Proof. Given $\{\pi_k\}_{k=1}^K$ and a set of minimum rate constraints $\{R_k^{\mathrm{avg}} \geq R_{k,\min}\}_{k=1}^K$, let point Y be the optimal solution in $\mathcal{S}_c \subseteq \mathcal{S}$. Assume Y is not on the boundary of $\mathcal{S}$. Let Y' be the intersection of the line OY with the boundary of $\mathcal{S}$. Since $OY' > OY$ and the points O, Y, and Y' are aligned, then the rates $R_{k,Y'} \geq R_{k,Y}$ for all k. Consequently, $\sum_{k=1}^K \pi_k R_{k,Y'} > \sum_{k=1}^K \pi_k R_{k,Y}$. Hence, Y' achieves the minimum rate constraints while leading to a weighted sum-rate greater than in Y, which contradicts the assumption that Y is the optimal solution. Therefore, the optimal solution in the case of rate constraints is on the boundary of the set $\mathcal{S} \cap \mathcal{S}_c$. ☐

3.4.3   Application to the Two-Users Rate Region

The case with $K = 2$ is shown in Fig. 3.1 that illustrates the two-user rate region for weighted uplink ergodic sum-rate. Point A corresponds to $(\pi_1 = 1, \pi_2 = 0)$, that is, all the subcarriers are allocated to user 1. The power of user 1 is distributed

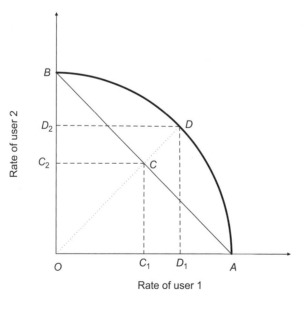

Figure 3.1   Rate region for two users with equal distances from the BS.

among the subcarriers according to the optimal allocation in (3.13). Similarly, point B corresponds to $(\pi_1 = 0, \pi_2 = 1)$.

Point C corresponds to symmetric time-sharing between the solutions in A and B, that is, allocating all the subcarriers to user 1 for 50% of the time and to user 2 for 50% of the time.

Point D corresponds to $(\pi_1 = 1/2, \pi_2 = 1/2)$, that is, both users are treated with equal priority. Hence, subcarriers are allocated to users according to (3.15) and the power of each user is subdivided by water-filling over the allocated subcarriers. Naturally, point D achieves a better performance than C for both users ($OD_1 \geq OC_1$ and $OD_2 \geq OC_2$) since it is obtained by optimal allocation of the subcarriers given the channel knowledge, whereas in point C subcarrier allocation is done without taking the channel knowledge into account. This statement can be mathematically expressed as

$$\max_{\mathbf{A},\mathbf{P}} \mathbb{E}_g \left\{ \sum_{k=1}^{2} \frac{1}{2} \sum_{i=1}^{N} \alpha_{k,i} \log_2(1 + P_{k,i} g_{k,i}) \right\} \geq$$
$$\max_{\mathbf{P}} \mathbb{E}_g \left\{ \frac{1}{2} \left(\sum_{i=1}^{N} \log_2(1 + P_{1,i} g_{1,i}) + \sum_{i=1}^{N} \log_2(1 + P_{2,i} g_{2,i}) \right) \right\} \tag{3.35}$$

It should be noted that, when the two users are equidistant from the BS, we have $OA = OB$, $OC_1 = OC_2$, and $OD_1 = OD_2$.

Fig. 3.2 shows the effect of rate constraints on reducing the achievable rate region. The per-user rate constraints $R_{1,\min}$ and $R_{2,\min}$ are represented by two lines perpendicular to the axes R_1 and R_2, respectively. The region S_c is shown hashed in the figure. In the upper left part, the rate constraints are set too high and hence there is no intersection and no solution satisfying the rate constraints can be reached ($S_c = \varnothing$). The upper right part shows a scenario where the same optimal solution is achievable in the presence and absence of rate constraints. The lower part of the figure shows two scenarios where the optimal solution moved from D to D' due to the rate constraints, according to Theorems 3.3 and 3.4. It is remarkable that in the two users case, the solution moved to the point not only nearest in terms of D_π, but also nearest in terms of the Euclidean distance D_E. This fact is proven in the following theorem.

Theorem 3.5. *In the two-users case, the solution maximizing the weighted sum-rate with rate constraints is the point on the boundary of $S \cap S_c$ having the minimum Euclidean distance to the solution maximizing the weighted sum-rate without rate constraints.*

Proof. Given $\{\pi_k\}_{k=1}^{K}$, let point X be the optimal solution in S. Furthermore, given a set of minimum rate constraints $\{R_k^{\text{avg}} \geq R_{k,\min}\}_{k=1}^{K}$, let point Y be the optimal solution in S_c. It is known from Theorem 3.4 that Y is on the boundary of $S \cap S_c$, and from Theorem 3.3 that $D_\pi(X, Y) = \min_{Y' \in S_c} D_\pi(X, Y')$. If $S_c \neq \varnothing$, then $\exists k$ such

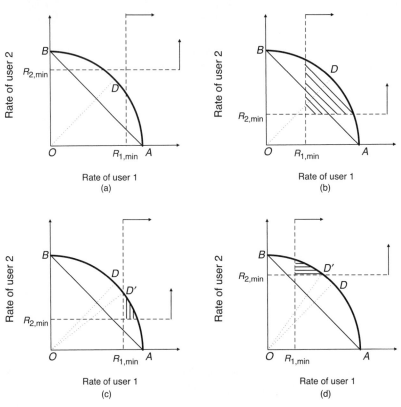

Figure 3.2   Rate region for two users with minimum rate constraints: (a) No solution satisfying the rate constraints is achievable; (b) the optimal solution is the same with and without rate constraints; (c) the rate constraints move the optimal solution from D to D' thus increasing the rate of user 1; (d) the rate constraints move the optimal solution from D to D' thus increasing the rate of user 2.

that $R_{k,X} \geq R_{k,\min}$. Without loss of generality, $R_{k,X}$ and $R_{k,Y}$ are sorted in ascending order. Hence,

$$R_{1,Y} \geq R_{1,\min} > R_{1,X}$$

Since X is the optimal solution in the absence of rate constraints, then

$$\pi_1 R_{1,X} + \pi_2 R_{2,X} > \pi_1 R_{1,Y} + \pi_2 R_{2,Y}$$

$$D_\pi(X, Y) = \pi_1(R_{1,X} - R_{1,Y}) + \pi_2(R_{2,X} - R_{2,Y}) > 0$$

Thus,

$$0 < \pi_1(R_{1,Y} - R_{1,X}) < \pi_2(R_{2,X} - R_{2,Y})$$

Since the first term in the summation $(R_{1,X} - R_{1,Y})$ is negative, then the second $(R_{2,X} - R_{2,Y})$ is definitely positive. Furthermore, since $Y = \arg \max_{Y' \in \mathcal{S}_c} (\pi_1 R_{1,Y'} +$

$\pi_2 R_{2,Y'}$), then $R_{1,Y}$ should be as close as possible to $R_{1,X}$ and $R_{2,Y}$ as close as possible to $R_{2,X}$. Consequently,

$$R_{1,Y} = R_{1,\min} \quad \text{and} \quad Y = \mathrm{argmin}_{Y' \in \mathcal{S}_c, R_{1,Y'} = R_{1,\min}} (R_{2,X} - R_{2,Y'})$$

Thus,

$$D_E(X, Y) = \min_{Y' \in \mathcal{S}_c} \sqrt{(R_{1,X} - R_{1,Y'})^2 + (R_{2,X} - R_{2,Y'})^2}$$

$\square$

3.5 RESULTS AND DISCUSSION

In this section, numerical results obtained by implementing the theoretical derivations of the previous sections are presented.

3.5.1 Simulation Parameters

For the simulations presented in this section, the considered channel has Rayleigh distributed fading. As a result, the channel gain of each user is exponentially distributed with a mean equal to the path gain representing propagation loss. Propagation loss is modeled using the path loss model given as

$$L_P = \kappa d_k^{-\upsilon} \tag{3.36}$$

where κ is the path loss constant chosen to be -128.1 dB, d_k is the distance in km from user k to the BS, and υ is the path loss exponent, set to a value of 3.76 that is typical for urban environments [2]. The simulation model consists of a single cell having a radius of 1 km, with the BS placed at its center and equipped with an omnidirectional antenna, and $K = 2$ users with a distance less than 1 km from the BS. The total bandwidth is $B = 5$ MHz divided into $N = 16$ subcarriers. The maximum transmission power per user is $P_k = 125$ mW for all k.

3.5.2 Multiplier Calculation and Convergence

Knowing the fading probability distribution function, enough samples $g_{k,i}$ can be generated offline to be used in the iterations of (3.19) and (3.20). The obtained λ_ks and μ_ks can then be used in instantaneous resource allocation as long as the fading distribution remains unchanged. In case offline training is not possible, convergence can be obtained online by using running averages over the instantaneous fading realizations. It should be noted that, with ergodicity, channel samples help in computing the Lagrangian parameters then using them without recomputation as long as the channel probability distribution function is the same. Conversely, recomputation of the Lagrangian parameters must be done with instantaneous scheduling at each scheduling instant. Hence, although channel samples are still needed at the scheduling instants, ergodicity leads to the fact that the λ_ks and μ_ks, computed once, ensure on average the optimal performance that can be obtained with instantaneous scheduling, where

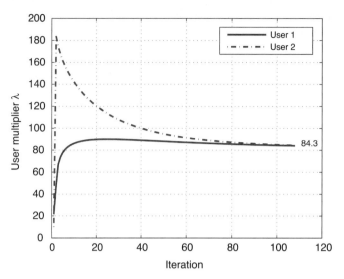

Figure 3.3 Evolution of the multiplier λ for the two users, equal distance case $(d_1 = d_2 = 200$ m).

the λ_ks and μ_ks must be computed at each scheduling instant using subgradient itera-tions. Thus, ergodic rate maximization reduces the scheduling load at the base station, avoids scheduling delays due to subgradient iterations at each scheduling instant, and ensures at the same time an optimal performance on the long-term average.

In order to illustrate the convergence of the Lagrangian multipliers, a system with two users having equally weighted rates and $R_{1,\min} = R_{2,\min} = 0$ is considered. Figs 3.3 and 3.4 show two different simulation configurations, one with the users

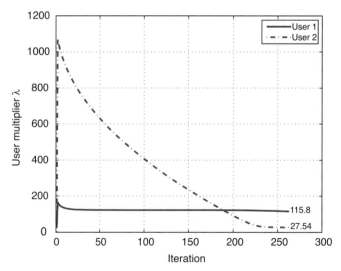

Figure 3.4 Evolution of the multiplier λ for the two users, different distance case $(d_1 = 200$ m, $d_2 = 400$ m).

placed at the same distance from the BS, and the other with one user placed at double the other user's distance from the BS. The initial multiplier values were selected as $\lambda_1^{(0)} = 10$ and $\lambda_2^{(0)} = 1$. This distinction was intentional in order to show that convergence is reached regardless of the initial value.

For the equal distance case, the values of λ_1 and λ_2 converge to the same result, which indicates that both users achieve the same rate on the long-term average, as expected. In the case of the different distance scenario, starting from the same initial multiplier vector, λ_1 converged to a value much larger than λ_2. Looking at λ as a power cost, it can be noted that user 1 is penalized more in order to allow some access for user 2 that has a lower multiplier value, although user 1 still gets most of the resources. From (3.12), it can be clearly seen that as the value of λ increases, that of the link metric Φ decreases. However, this increase in the value of λ_1 did not considerably change the outcome of the allocation due to the better channel conditions that favor user 1 over user 2, as will be seen in Section 3.5.4. It can also be noted that the number of iterations for convergence is different. This depends on the chosen values for the step sizes and for the initial multiplier values.

However, the convergence of the multipliers is not as critical as the convergence of the rate. Fig. 3.5 shows the convergence of the sum-rate when the computations are performed online by using instantaneous channel samples $g_{k,i}$. It can be seen in Fig. 3.5 that the total rate increases from 13.23 Mbps when 10 samples are available at the scheduler to 13.75 Mbps when 10,000 samples are available. It can also be noticed that after 3600 samples the sum-rate value does not vary by a considerable amount and after 8700 samples, the value of the sum-rate does not change anymore. Thus, with 3600 samples, 99% of the optimal sum-rate is achieved and 96% is reached with only 10 samples. Hence, considering 1 ms scheduling intervals, practical convergence occurred after only 10 ms.

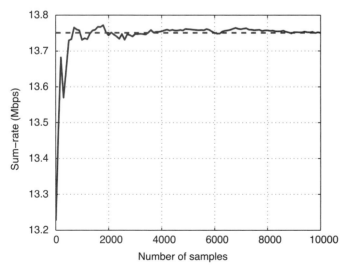

Figure 3.5 Evolution of the sum-rate with the number of samples. The dashed line marks the value at which convergence occurs.

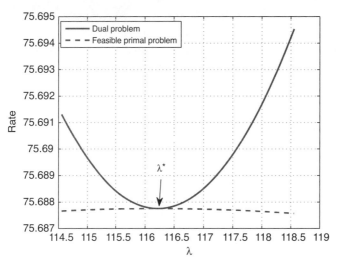

Figure 3.6 Primal and dual values as a function of λ.

3.5.3 Duality Gap Results

In order to be able to plot the duality gap, a scenario consisting of two users with equal distances to the BS and having equally weighted rates with $R_{1,\min} = R_{2,\min} = 0$ is considered.

Fig. 3.6 shows the evolution of the dual and primal values as a function of λ. The duality gap is then the difference between the dual and primal solution for each value of λ. This figure clearly demonstrates the convexity of the dual function and the concavity of the primal function. It can also be noted that as the value of λ approaches the optimal value λ^* (note that $\lambda_1^* = \lambda_2^*$), the duality gap goes to 0 and then it increases again. Hence, when the total power achieved by one user is equal to its maximum allowable power, the obtained optimal vector of multipliers achieves zero duality gap.

3.5.4 Sum-Rate Results

To illustrate the effect of weights on the resource allocation, a scenario of two users placed at equal distance from the BS is considered. The weight of user 1 is set to π_1 between 0 and 1, while the weight of user 2 is set to $\pi_2 = 1 - \pi_1$. No rate constraints are considered ($R_{1,\min} = R_{2,\min} = 0$).

Fig. 3.7 shows the results of the simulation. As $\pi_1 < 0.5$, the rate of user 2 is greater than that of user 1. As expected, the break-even point is when the weights of both users are equal, that is, $\pi_1 = \pi_2 = 0.5$. In this case, the rates of user 1 and user 2 are equal since the weights are not playing any role in favoring one user over the other. Fig. 3.7 shows that each of the two users achieves a rate of 9.15 Mbps when the weights are equal. Furthermore, it should be noted that the maximum sum-rate is achieved when the weights are equal. Plotting the rate of user 2 versus that of user 1 would lead to a figure having the same form as Fig. 3.1 with $OA = OB = 11.27$ Mbps, $OC_1 = OC_2 = 5.635$ Mbps, and $OD_1 = OD_2 = 9.15$ Mbps.

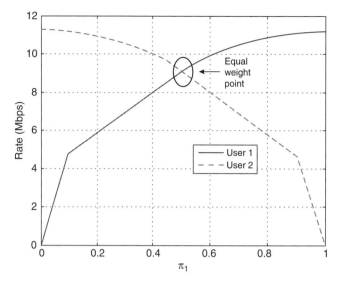

Figure 3.7   Evolution of the users' rates as a function of the weights. The users are located at $d_1 = d_2 = 200$ m from the BS.

To investigate the effect of the weights on the users' rates and therefore on the fairness in the system, the scenario of two users, one placed at 200 m from the BS, and the other at 400 m from the BS, is considered. The weights of these users are varied between 0 and 1. The results are shown in Table 3.1.

The weight of user 2, which is placed farther than user 1 from the BS, has allowed that user to be allocated some subcarriers and therefore achieve some rate particularly when its corresponding weight is high relative to that of user 1. However, as the weight of user 2 tends to 0, it provides less advantage to the user, since a low weight corresponds to imposing a penalty on user 2 that already has unfavorable channel

TABLE 3.1   Effect of Weights on the Users' Rates ($d_1 = 200$ m, $d_2 = 400$ m)

(π_1, π_2)	$(\lambda_1^*, \lambda_2^*)$	(R_1, R_2) (Mbps)
(0, 1)	(0.0001, 45.3419)	(0, 2.85)
(0.1, 0.9)	(8.9883, 37.2735)	(9.10, 2.78)
(0.2, 0.8)	(19.5470, 30.1559)	(9.79, 2.67)
(0.3, 0.7)	(31.1061, 23.6864)	(10.23, 2.44)
(0.4, 0.6)	(43.4735, 18.1772)	(10.66, 2.32)
(0.5, 0.5)	(56.2455, 13.6410)	(10.93, 2.09)
(0.6, 0.4)	(69.3622, 9.8178)	(11.09, 1.91)
(0.7, 0.3)	(82.6455, 6.6467)	(11.17, 1.74)
(0.8, 0.2)	(96.0789, 3.9782)	(11.22, 1.58)
(0.9, 0.1)	(109.1407, 1.7562)	(11.25, 1.41)
(1, 0)	(122.4817, 0.0001)	(11.27, 0)

TABLE 3.2 Effect of Rate Constraints on Rate Maximization ($d_1 = 200$ m, $d_2 = 400$ m)

(π_1, π_2)	$(\lambda_1^*, \lambda_2^*)$	(μ_1^*, μ_2^*)	(R_1, R_2) (Mbps)
(0, 1)	(37.2124, 44.3201)	(0.0847, 0)	(2.50, 2.82)
(0.1, 0.9)	(8.9018, 37.5858)	(0, 0)	(9.10, 2.78)
(0.2, 0.8)	(19.8292, 30.1082)	(0, 0)	(9.79, 2.67)
(0.3, 0.7)	(29.8916, 32.1086)	(0, 0.1871)	(10.21, 2.50)
(0.4, 0.6)	(40.7194, 39.9994)	(0, 0.5208)	(10.21, 2.50)
(0.5, 0.5)	(52.7813, 40.0876)	(0, 0.6742)	(10.21, 2.50)
(0.6, 0.4)	(62.8856, 51.2685)	(0, 1.0742)	(10.21, 2.50)
(0.7, 0.3)	(73.0412, 57.1155)	(0, 1.3660)	(10.21, 2.50)
(0.8, 0.2)	(84.8819, 61.1053)	(0, 1.6085)	(10.21, 2.50)
(0.9, 0.1)	(95.7576, 65.0692)	(0, 1.8585)	(10.21, 2.50)
(1, 0)	(103.9174, 70.0120)	(0, 2.1436)	(10.21, 2.50)

conditions. When a user has a weight of 0, its rate does not have any contribution in the weighted maximization and hence the rate is set to zero in order to allocate all resources to the other user. Given that user 1 is closer to the BS than user 2, while a low weight decreases user 1's rate by almost 2.15 Mbps from the maximum achievable rate, its rate remains larger than that of user 2. These results show that giving higher weights to user 2 that is less favored by its channel conditions, allows it to enhance its performance compared to the equal weights scenario. This can be seen by comparing, in Table 3.1, the results with $(\pi_1 = 0.1, \pi_2 = 0.9)$ to those with $(\pi_1 = 0.5, \pi_2 = 0.5)$. Plotting the rate of user 2 versus that of user 1 would lead to a figure having the same form as Fig. 3.1 with $OA = 11.27$ Mbps, $OB = 2.85$ Mbps, $OC_1 = 5.635$ Mbps, $OC_2 = 1.425$ Mbps, $OD_1 = 10.93$ Mbps, and $OD_2 = 2.09$ Mbps.

Adding a minimum rate constraint $R_{1,min} = R_{2,min} = 2.5$ Mbps, the results of Table 3.1 evolve to those of Table 3.2. It is observed that user 2 achieves its minimum rate of 2.5 Mbps although it is twice as far from the BS than user 1. As a result, it can be noted that the addition of the user weights and rate constraints has improved the fairness in the system. Plotting the rate of user 2 versus that of user 1 would lead to a figure having the same form as Fig. 3.2d, where, for example, in the case $(\pi_1 = \pi_2 = 0.5)$, the optimal solution moved from $D(10.93, 2.09)$ to $D'(10.21, 2.50)$. For the two extreme cases where the weights are taken as $(\pi_1 = 0, \pi_2 = 1)$ or $(\pi_1 = 1, \pi_2 = 0)$, the rates of user 1 and user 2 were respectively equal to 2.5 Mbps. The constraint is enforced in these cases since it is imposed on the rate itself, not on the weighted rate. This corresponds to scenarios (c) and (d) of Fig. 3.2, where the optimal solution moved from B to D' and from A to D', respectively. Hence, the ergodic weighted sum-rate is maximized given that the power and rate constraints are satisfied.

Comparing the results of Table 3.2 to those of Table 3.1, it can be seen that the results of μ_1^* and μ_2^* in Table 3.2 are in line with the conclusions of Section 3.3.2 in that the parameter is set to zero when a user achieves a rate higher than the minimum rate constraint in order to have a zero duality gap. Otherwise, μ^* converges to a value that leads to meeting the rate constraint with equality, according to Theorem 3.5.

3.6 SUMMARY

The problem of weighted ergodic sum-rate maximization in the uplink of single cell OFDMA systems with power and rate constraints was investigated. Starting from a convex optimization formulation of the problem, an analytical solution was derived and the duality gap was analyzed. The characteristics of the achievable rate region were determined for an arbitrary number of users, and the special case of the two users rate region was studied. Numerical results were presented via simulations, and the computation of the Lagrangian parameters were shown to converge to the final solution within a limited duration.

This chapter addressed ergodic sum-rate maximization with continuous rates, where the Shannon capacity formula, $\log(1 + \text{SNR})$, is used for the rate expression. This corresponds to a scenario where an infinite number of rates can be achieved by each user. In Chapter 4, the problem of weighted ergodic sum-rate maximization in the uplink of single cell OFDMA systems with discrete rates is considered. This problem corresponds to the practical scenario where a finite number of rates is achieved via a fixed number of modulation and coding schemes.

CHAPTER *4*

ERGODIC SUM-RATE MAXIMIZATION WITH DISCRETE RATES

In Chapter 3, the solution of weighted ergodic sum-rate maximization subject to per-user power and rate constraints within a single cell was presented for the continuous rates case. In this chapter, the solution of the same problem is presented for the discrete rates case. The chapter is structured as follows. Section 4.1 presents a review of the related literature and outlines the main ideas investigated in this chapter. The problem formulation is given in Section 4.2. The solution is described in Section 4.3. Numerical results are presented and discussed in Section 4.4. Concluding remarks follow in Section 4.5.

4.1 BACKGROUND

The Shannon capacity formula for Gaussian channels, $\log_2(1 + P_{k,i}g_{k,i})$, was adopted in Chapter 3 to solve the problem of weighted ergodic sum-rate maximization with continuous rates. This capacity expression is based on the assumption of infinite length codewords generated according to a normal distribution [38]. Although this formula is widely used in the literature, for example [19, 20, 24, 26, 27, 33, 47, 95–97], only a discrete set of rates are achievable in a practical system, due to a fixed number of modulation and coding schemes (MCS). Hence, it would be interesting to formulate and solve the problem of ergodic sum-rate maximization with discrete rates. In Ref. [26], this problem is investigated for the downlink in addition to the continuous rates scenario. This chapter targets the investigation of the problem with discrete rates in the context of uplink ergodic sum-rate maximization. The main topics discussed in this chapter are summarized as follows:

1. Formulating and solving the ergodic weighted sum-rate maximization problem subject to per user power and rate constraints in the uplink of OFDMA systems for the discrete rates case. In Ref. [26], this problem is investigated for the downlink in addition to the continuous rates scenario. In this chapter, the practical problem with discrete rates is investigated in the context of uplink ergodic

Resource Allocation in Uplink OFDMA Wireless Systems: Optimal Solutions and Practical Implementations, Elias E. Yaacoub and Zaher Dawy.
© 2012 by the Institute of Electrical and Electronics Engineers, Inc. Published 2012 by John Wiley & Sons, Inc.

sum-rate maximization. Furthermore, minimum rate constraints are considered in the problem in addition to the power constraints.

2. Analyzing the duality gap of the proposed solution and showing that a zero duality gap is reached when the optimal dual solution is derived.

3. Investigating the performance differences between the optimal solution with continuous rates and the optimal solution with discrete rates. The effect of the target bit error rate (BER), the impact of an increased number of modulation schemes, and the role of channel coding in reducing the performance gap are studied.

4.2 PROBLEM FORMULATION

As in Chapter 3, a single cell uplink OFDMA system is considered, with K users and N subcarriers to be allocated. The CNR is given by (3.1). The exclusivity of subcarrier allocation is given by (3.2). The power constraint is expressed by (3.3).

However, in the discrete rates case, the total throughput of user k is defined as follows:

$$R_k = \sum_{i=1}^{N} \alpha_{k,i} R_{k,i}^d(P_{k,i}, g_{k,i}) \tag{4.1}$$

where $R_{k,i}^d$ is the discrete rate of user k over subcarrier i. Conversely to continuous rates, which can take any non-negative real value according to the Shannon capacity formula $\log_2(1 + P_{k,i}g_{k,i})$, discrete rates represent the quantized bit rates achievable in a practical system as follows:

$$R_{k,i}^d(P_{k,i}, g_{k,i}) = \begin{cases} r_0, & \eta_0 \le P_{k,i}g_{k,i} < \eta_1 \\ r_1, & \eta_1 \le P_{k,i}g_{k,i} < \eta_2 \\ r_2, & \eta_2 \le P_{k,i}g_{k,i} < \eta_3 \\ \vdots & \vdots \\ r_{L-1}, & \eta_{L-1} \le P_{k,i}g_{k,i} < \eta_L \end{cases} \tag{4.2}$$

where η_l represents the SNR target in order to achieve the rate r_l with a predefined BER. Note that in the limit, it can be written that $r_0 = 0$, $\eta_0 = 0$, and $\eta_L = \infty$. Examples of rates satisfying these conditions are shown in Fig. 4.1.
Consequently, the total system throughput is given by

$$R(\mathbf{A}, \mathbf{P}) = \sum_{k=1}^{K} \sum_{i=1}^{N} \alpha_{k,i} R_{k,i}^d(P_{k,i}, g_{k,i}) \tag{4.3}$$

where $\mathbf{A}$ is a $K \times N$ matrix of channel allocation indices, $\alpha_{k,i}$, and $\mathbf{P}$ is a $K \times N$ matrix of allocated powers $P_{k,i}$.

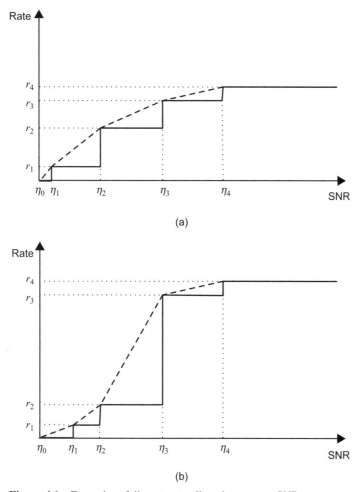

Figure 4.1 Examples of discrete rate allocations versus SNR.

Although the rate functions in Fig. 4.1 are not concave, they can be shown to be quasiconcave. In fact, the t-superlevel set of $R^d_{k,i}(\eta_{k,i})$ is defined as $C_t = \{\eta_{k,i} \in \mathcal{R} | R^d_{k,i}(\eta_{k,i}) \geq t\}$, with $\mathcal{R}$ the set of real numbers. Since the rates are increasing with the SNR, the rate function is quasiconcave because each of its superlevel sets is convex [34]. In addition, letting $\eta_{k,i} = P_{k,i} g_{k,i}$, the rate function is uniformly quasiconcave since if $\eta_{k,i} \leq \eta'_{k,i}$ then $\min(R^d_{k,i}(\eta_{k,i}), R^d_{k,i}(\eta'_{k,i})) = R^d_{k,i}(\eta_{k,i})$ for all k and i [108]. Although the sum of quasiconcave functions is not necessarily quasiconcave, the sum of uniformly quasiconcave functions, on the same set, is quasiconcave [108]. A quasiconcave function can always be represented by a family of concave functions, one for each superlevel set such that [34] $R^d_{k,i}(\eta_{k,i}) \geq t \Leftrightarrow \phi_t(\eta_{k,i}) \geq 0$ and $\phi_s(\eta_{k,i}) \geq \phi_t(\eta_{k,i})$ when $s \geq t$. In Ref. [34], a method to solve the quasiconcave optimization problem is presented. It consists of using a sequence of convex feasibility problems and the method of bisection. Starting from an initial interval known to contain the optimal solution, in each step, the interval is narrowed by half of its width, until the width becomes smaller than an arbitrary value ϵ. The quasiconcavity property will allow us to prove in Section 4.3.1 that the dual solution is achievable with zero duality gap, and hence that the optimal solution of the primary problem can be reached via the optimal solution of the dual.

The problem of the maximization of the weighted ergodic sum-rate can be formulated as follows:

$$\max_{\mathbf{A},\mathbf{P}} \mathbb{E}_g\left\{ \sum_{k=1}^{K} \pi_k \sum_{i=1}^{N} \alpha_{k,i} R^d_{k,i}(P_{k,i}, g_{k,i}) \right\}$$

subject to

$$\mathbb{E}_g\left\{ \sum_{i=1}^{N} P_{k,i} \right\} \leq P_{k,\max}, \quad \text{for all } k \tag{4.4}$$

$$\mathbb{E}_g\left\{ \sum_{i=1}^{N} \alpha_{k,i} R^d_{k,i}(P_{k,i}, g_{k,i}) \right\} \geq R_{k,\min}, \quad \text{for all } k$$

where $\mathbb{E}\{.\}$, π_k, $P_{k,\max}$, and $R_{k,\min}$ represent the same quantities as in Chapter 3.

4.3 PROBLEM SOLUTION

To solve the problem in (4.4), the problem is first reformulated into an equivalent form that leads to the same solution but is easier to solve. By letting

$$f_{k,i} = \alpha_{k,i} P_{k,i} \tag{4.5}$$

It can be easily shown that

$$\begin{aligned}
\alpha_{k,i} R^d_{k,i}(P_{k,i}, g_{k,i}) &= \alpha_{k,i} R^d_{k,i}(\alpha_{k,i} P_{k,i}, g_{k,i}) \\
&= \alpha_{k,i} R^d_{k,i}(f_{k,i}, g_{k,i})
\end{aligned} \tag{4.6}$$

since $\alpha_{k,i}$ can only take 0 or 1 values.

Consequently, the problem in (4.4) is equivalent to

$$\max_{\mathbf{A},\mathbf{F}} \mathbb{E}_g\left\{ \sum_{k=1}^{K} \pi_k \sum_{i=1}^{N} \alpha_{k,i} R^d_{k,i}(f_{k,i}, g_{k,i}) \right\}$$

subject to

$$\mathbb{E}_g\left\{ \sum_{i=1}^{N} f_{k,i} \right\} \leq P_{k,\max}, \quad \text{for all } k \tag{4.7}$$

$$\mathbb{E}_g\left\{ \sum_{i=1}^{N} \alpha_{k,i} R^d_{k,i}(f_{k,i}, g_{k,i}) \right\} \geq R_{k,\min}, \quad \text{for all } k$$

since finding an optimal allocation $(\alpha_{k,i}^*, \alpha_{k,i}^* P_{k,i}^*)$ in (4.7) is equivalent to finding $(\alpha_{k,i}^*, P_{k,i}^*)$ in (4.4).

In order to reach the solution of the problem defined in (4.7), the Lagrangian is defined. It entails a vector of Lagrange multipliers λ corresponding to the power constraint and μ corresponding to the rate constraint.

$$
\begin{aligned}
L(\mathbf{A}, \mathbf{F}, \lambda, \mu) = \mathbb{E}_g\Bigg\{ &\sum_{k=1}^{K} \pi_k \sum_{i=1}^{N} \alpha_{k,i} R_{k,i}^d(f_{k,i}, g_{k,i}) \Bigg\} \\
&- \sum_{k=1}^{K} \lambda_k \left(\mathbb{E}_g\Bigg\{ \sum_{i=1}^{N} f_{k,i} \Bigg\} - P_{k,\max} \right) \\
&- \sum_{k=1}^{K} \mu_k \left(R_{k,\min} - \mathbb{E}_g\Bigg\{ \sum_{i=1}^{N} \alpha_{k,i} R_{k,i}^d(f_{k,i}, g_{k,i}) \Bigg\} \right)
\end{aligned}
\tag{4.8}
$$

The Lagrangian dual function is then given by

$$
D(\lambda, \mu) = \max_{\mathbf{A}, \mathbf{F}} L(\mathbf{A}, \mathbf{F}, \lambda, \mu)
\tag{4.9}
$$

The optimization dual problem is given by

$$
\min_{\lambda \geq 0, \mu \geq 0} D(\lambda, \mu)
\tag{4.10}
$$

The Lagrangian dual can be rewritten, after interchanging the expectation and the summation, as follows:

$$
\begin{aligned}
D(\lambda, \mu) &= \max_{\mathbf{A}, \mathbf{F}} \sum_{k=1}^{K} \sum_{i=1}^{N} \mathbb{E}_g\Big\{ (\pi_k + \mu_k) - \lambda_k f_{k,i} \Big\} \alpha_{k,i} R_{k,i}^d(f_{k,i}, g_{k,i}) \\
&\quad + \sum_{k=1}^{K} \lambda_k P_{k,\max} - \sum_{k=1}^{K} \mu_k R_{k,\min} \\
&= \max_{\mathbf{A}} \sum_{k=1}^{K} \sum_{i=1}^{N} \max_{\mathbf{P}} \mathbb{E}_g\Big\{ \alpha_{k,i} \left((\pi_k + \mu_k) R_{k,i}^d(P_{k,i}, g_{k,i}) - \lambda_k P_{k,i} \right) \Big\} \\
&\quad + \sum_{k=1}^{K} \lambda_k P_{k,\max} - \sum_{k=1}^{K} \mu_k R_{k,\min} \\
&= \max_{\mathbf{A}} \sum_{k=1}^{K} \sum_{i=1}^{N} \max_{\mathbf{P}} \mathbb{E}_g\Big\{ \alpha_{k,i} \Phi(P_{k,i}) \Big\} + \sum_{k=1}^{K} \lambda_k P_{k,\max} - \sum_{k=1}^{K} \mu_k R_{k,\min}
\end{aligned}
\tag{4.11}
$$

with $\Phi(P_{k,i})$ defined as follows:

$$
\Phi(P_{k,i}) = (\pi_k + \mu_k) R_{k,i}^d(P_{k,i}, g_{k,i}) - \lambda_k P_{k,i}
\tag{4.12}
$$

Consequently,

$$
\begin{aligned}
\Phi(P_{k,i}) &= (\pi_k + \mu_k)R_{k,i}^d(P_{k,i}, g_{k,i}) - \lambda_k P_{k,i} \\
&= (\pi_k + \mu_k)r_l - \lambda_k P_{k,i} \\
&\le (\pi_k + \mu_k)r_l - \lambda_k \frac{\eta_l}{g_{k,i}}
\end{aligned}
\tag{4.13}
$$

Due to the discrete nature of the rates, there are L candidate power allocation functions

$$
P_{k,i} \in \left\{ \frac{\eta_0}{g_{k,i}}, \frac{\eta_1}{g_{k,i}}, \ldots, \frac{\eta_{L-1}}{g_{k,i}} \right\}
\tag{4.14}
$$

from which the one maximizing (4.13) should be chosen. Hence,

$$
l_{k,i}^* = \arg\max_l (\pi_k + \mu_k)r_l - \lambda_k \frac{\eta_l}{g_{k,i}}
\tag{4.15}
$$

Then,

$$
R_{k,i}^* = r_{l_{k,i}^*}
\tag{4.16}
$$

and

$$
P_{k,i}^* = \frac{\eta_{l_{k,i}^*}}{g_{k,i}}
\tag{4.17}
$$

Due to the exclusivity of subcarrier allocation, (4.12) is maximized by allocating subcarrier i to user k_i^* such that

$$
\begin{aligned}
k_i^* &= \arg\max_k (\pi_k + \mu_k)r_{l_{k,i}^*} - \lambda_k P_{k,i}^* \\
&= \arg\max_k (\pi_k + \mu_k)r_{l_{k,i}^*} - \lambda_k \frac{\eta_{l_{k,i}^*}}{g_{k,i}}
\end{aligned}
\tag{4.18}
$$

Hence, the optimal subcarrier allocation is given by

$$
\alpha_{k,i}^* = \begin{cases} 1, & \text{if } k = k_i^* \\ 0, & \text{otherwise} \end{cases}
\tag{4.19}
$$

and the optimal power allocation is given by

$$
f_{k,i}^* = \alpha_{k,i}^* P_{k,i}^* = \begin{cases} \dfrac{\eta_{l_{k,i}^*}}{g_{k,i}}, & \text{if } k = k_i^* \\ 0, & \text{otherwise} \end{cases}
\tag{4.20}
$$

In (4.15) and (4.18), λ_k acts as a power price that penalizes the user as the transmitted power increases, whereas μ_k acts as a rate reward that rewards the user when its achieved rate increases.

To complete the solution, the vector of geometric multipliers λ associated with the power constraint and the vector of geometric multipliers μ associated with the rate constraint should be determined. Determining the multipliers can be done using an iterative subgradient method with each iteration on the multipliers λ_k and μ_k given by

$$\lambda_k^{n+1} = \left[\lambda_k^n - \delta_n G_{\lambda_k}^n\right]^+$$
$$\mu_k^{n+1} = \left[\mu_k^n - \delta_n G_{\mu_k}^n\right]^+ \tag{4.21}$$

where the superscript n denotes the index of the iteration and G_λ^n and G_μ^n denote the subgradients, which are taken as

$$G_{\lambda_k}^n = P_{k,\max} - \mathbb{E}_g\left\{\sum_{i=1}^{N} P_{k,i}^* \alpha_{k,i}^*\right\}$$
$$G_{\mu_k}^n = \mathbb{E}_g\left\{\sum_{i=1}^{N} \alpha_{k,i}^* R_{k,i}^d(\alpha_{k,i}^* P_{k,i}^*, g_{k,i})\right\} - R_{k,\min} \tag{4.22}$$

The step size δ_n is taken of the form $\delta_n = a/\sqrt{n}$, where a is a positive constant. This chosen step size is guaranteed to lead to convergence since it obeys the nonsummable diminishing rule [35]:

$$\lim_{n \to \infty} \delta_n = 0, \qquad \sum_{n=1}^{\infty} \delta_n = \infty. \tag{4.23}$$

Knowing the fading probability distribution function, enough samples can be generated offline to be used in the iterations of (4.21) and (4.22). The obtained λ_ks and μ_ks can then be used in instantaneous resource allocation without recomputation as long as the fading distribution remains unchanged. In case offline training is not possible, convergence can be obtained online by using running averages over the instantaneous fading realizations. In this case, it is not required that the BS knows the fading pdf, as long as it is aware of the instantaneous fading realizations. Such an online training approach was shown in Ref. [36] to lead to the same solution as the offline case in the context of ad hoc peer-to-peer cognitive radio networks. However, in instantaneous sum-rate maximization corresponding to (4.4) without the expectation, the λ_ks and μ_ks have to be computed at each scheduling instant. This requires subgradient iterations in order to compute $2K$ parameters at every transmission time interval (TTI). Ergodic sum-rate maximization allows to avoid this overhead by taking advantage of the time dimension in addition to frequency and multiuser diversity dimensions. It should be noted that in both ergodic and instantaneous scheduling, the rate, subcarrier and power allocations are performed at every fading state. The difference is in computing the λ_ks and μ_ks: With ergodic scheduling, the optimal rate, subcarrier, and power allocations are computed at each fading state using the current power prices (λ_ks) and rate rewards (μ_ks); at a slower time scale, the power prices and rate rewards are adjusted to meet the average power constraints, similarly to the approach described in Ref. [37]. However, with instantaneous scheduling, the power prices and rate rewards are adjusted at every fading state.

4.3.1   Duality Gap Analysis

In this section, the solution derived in Section 4.3 is shown to be unique and corresponding to zero duality gap. Consequently, the solution of the dual problem can be directly mapped to the solution of the primal utility maximization problem.

The solution in (4.18) may not be unique if, for a given subcarrier i, the power prices and fading states of at least two users k_1 and k_2 satisfy $\frac{\lambda_{k_1}}{g_{k_1,i}} = \frac{\lambda_{k_2}}{g_{k_2,i}}$. But since the fading distribution has a continuous density, this situation occurs with probability 0 [37]. Hence, the optimal solution is unique. However, when only a finite set of fading states is available for each user, ties occur with a non-negligible probability, and finding the optimal solution requires time-sharing between users. This situation is investigated in Ref. [97] for the continuous rates scenario, and efficient suboptimal techniques with exclusive subcarrier allocation (without time-sharing) are proposed. In this book, the fading is assumed to have a continuous density and thus is absolutely continuous with respect to Lebesgue measure, as in Ref. [37].

Using the obtained power and subcarrier allocation, the dual solution can be formulated as follows:

$$
\begin{aligned}
d^* = \sum_{k=1}^{K} \sum_{i=1}^{N} \mathbb{E}_g \Big\{ (\pi_k + \mu_k^*) \alpha_{k,i}^* R_{k,i}^d(P_{k,i}^*, g_{k,i}) - \lambda_k^* \alpha_{k,i}^* P_{k,i}^* \Big\} \\
+ \sum_{k=1}^{K} \lambda_k^* P_{k,\max} - \sum_{k=1}^{K} \mu_k^* R_{k,\min}
\end{aligned}
\tag{4.24}
$$

The average power constraint states that the average power used by each user should be less than the user's target maximum power $P_{k,\max}$. The average power of the user is denoted by

$$
P_k^{\text{avg}}(\lambda_k^*, \mu_k^*) = \mathbb{E}_g \Big\{ \sum_{i=1}^{N} \alpha_{k,i}^* P_{k,i}^* \Big\}
\tag{4.25}
$$

The minimum rate constraint states that the average rate of each user should be greater than the user's target minimum rate $R_{k,\min}$. The average rate of user k is denoted by

$$
R_k^{\text{avg}}(\lambda_k^*, \mu_k^*) = \mathbb{E}_g \Big\{ \sum_{i=1}^{N} \alpha_{k,i}^* R_{k,i}^d(P_{k,i}^*, g_{k,i}) \Big\}
\tag{4.26}
$$

The dual solution can therefore be rewritten as

$$
\begin{aligned}
d^* = \sum_{k=1}^{K} \sum_{i=1}^{N} \mathbb{E}_g \Big\{ \pi_k \alpha_{k,i}^* R_{k,i}^d(P_{k,i}^*, g_{k,i}) \Big\} \\
+ \sum_{k=1}^{K} \lambda_k^* \Big(P_{k,\max} - P_k^{\text{avg}}(\lambda_k^*, \mu_k^*) \Big) \\
+ \sum_{k=1}^{K} \mu_k^* \Big(R_k^{\text{avg}}(\lambda_k^*, \mu_k^*) - R_{k,\min} \Big)
\end{aligned}
\tag{4.27}
$$

The primal solution can be written as

$$p^* = \mathbb{E}_g \left\{ \sum_{k=1}^{K} \sum_{i=1}^{N} \pi_k \alpha_{k,i}^* R_{k,i}^d(P_{k,i}^*, g_{k,i}) \right\} \tag{4.28}$$

From the weak duality theorem [35], given a problem with a dual solution d^* and a primal solution p^*, then

$$p^* \leq d^* \tag{4.29}$$

The duality gap is given by

$$d^* - p^* =$$

$$\sum_{k=1}^{K} \sum_{i=1}^{N} \pi_k \mathbb{E}_g \left\{ \alpha_{k,i}^* R_{k,i}^d(P_{k,i}^*, g_{k,i}) \right\} + \sum_{k=1}^{K} \lambda_k^* \left(P_{k,\max} - P_k^{\mathrm{avg}}(\lambda_k^*, \mu_k^*) \right)$$

$$- \mathbb{E}_g \left\{ \sum_{k=1}^{K} \sum_{i=1}^{N} \pi_k \alpha_{k,i}^* R_{k,i}^d(P_{k,i}^*, g_{k,i}) \right\} + \sum_{k=1}^{K} \mu_k^* \left(R_k^{\mathrm{avg}}(\lambda_k^*, \mu_k^*) - R_{k,\min} \right)$$

$$\tag{4.30}$$

From this equation, it can be easily shown that the duality gap is zero whenever

$$\lambda_k^* \left(P_{k,\max} - P_k^{\mathrm{avg}}(\lambda_k^*, \mu_k^*) \right) = 0, \quad \text{for all } k \tag{4.31}$$

and

$$\mu_k^* \left(R_k^{\mathrm{avg}}(\lambda_k^*, \mu_k^*) - R_{k,\min} \right) = 0, \quad \text{for all } k \tag{4.32}$$

To set (4.31) to zero, the condition $P_k^{\mathrm{avg}} = P_{k,\max}$ must be satisfied for all k. On the other hand, to satisfy (4.32), the condition $R_k^{\mathrm{avg}} = R_{k,\min}$ must be met only for users that have $R_k^{\mathrm{avg}} < R_{k,\min}$ in the optimal solution of the weighted ergodic sum-rate maximization problem without rate constraints, that is, when the rate constraints are not explicitly imposed. Users achieving $R_k^{\mathrm{avg}} > R_{k,\min}$ must have $\mu_k^* = 0$ in order to reach a zero duality gap, since forcing these users to achieve only the minimum rate leads to a suboptimal solution, due to reducing their contribution in the weighted ergodic sum-rate.

It should be noted that, although the primal function is not concave, it is uniformly quasiconcave and the primal solution can be found via a series of convex feasibility problems [34]. However, when solving via the dual solution instead, the situation is modeled in Fig. 4.2, showing the primal and dual values as a function of λ. A similar approach can be followed for the case of μ. In the case of Fig. 4.1a, connecting the edges of the staircase function leads to a concave function and hence to a zero duality gap solution with the primal value being the same as that of the initial staircase function. But the proposed approach can still be applied for the scenario of Fig. 4.1b. In fact, the primal optimal solution corresponds to any value in $[\lambda_a, \lambda_b]$. However, only the value at $\lambda^* \in [\lambda_a, \lambda_b]$ corresponds to zero duality gap. Since solving the dual problem leads to finding the solution at λ^*, and since any value in $[\lambda_a, \lambda_b]$

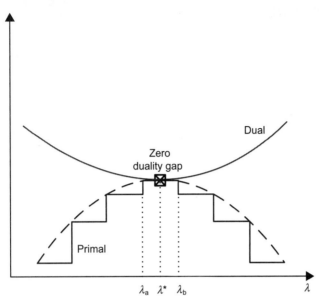

Figure 4.2 Duality gap.

corresponds to the same value of the optimal primal solution, then the dual solution leads to the optimal solution of the primal problem with zero duality gap.

4.3.2 Complexity Analysis

The optimal solution requires a complexity of $\mathcal{O}(INKL)$ in the Lagrangian parameters computation phase, with I the number of iterations. Afterwards, the complexity becomes $\mathcal{O}(NKL)$. However, it can be easily shown that (4.15) can be transformed into a table lookup operation, as proven in Ref. [26] for the downlink scenario. In this case, the complexity of the optimal solution at runtime becomes $\mathcal{O}(NK\ln(L))$. The optimal ergodic sum-rate maximization solution is still simpler to implement than the optimal instantaneous sum-rate maximization, which corresponds to (4.4) without the expectation. In fact, instantaneous sum-rate maximization requires a complexity of $\mathcal{O}(INKL)$ at each scheduling interval, since the Lagrangian parameters need to be computed at every scheduling instant in order to obtain the optimal performance.

4.4 RESULTS AND DISCUSSION

In this section, after presenting the simulation model, the optimal solution with discrete rates is compared to the optimal solution with continuous rates. Afterwards, the impact of the weights on the user rates is investigated.

4.4.1 Simulation Model

The fading is considered to have a Rayleigh distribution and hence the channel gain of each user is exponentially distributed with mean equal to the path gain representing propagation loss and given by $\kappa d_k^{-\upsilon}$, with κ a constant chosen to be -128.1 dB, d_k the distance in km from user k to the BS, and υ the path loss exponent, set to a value of 3.76. The simulation model consists of a single BS having an omnidirectional antenna. The rate is averaged over 10,000 TTIs, with the duration of a TTI being 1 ms. The total bandwidth considered is subdivided into $N = 16$ subcarriers. The maximum user transmit power is considered to be 125 mW.

4.4.2 Continuous Versus Discrete Rates

Fig. 4.3 shows the results of the optimal solution with continuous rates, compared to the optimal solution with discrete rates for a BER of 10^{-3} and 10^{-6}. The SNR thresholds for the discrete rates case are obtained from the BER approximation for Gray coded MQAM modulation without channel coding [109] given by

$$P_b \approx 0.2e^{\frac{-1.6P_{k,i}g_{k,i}}{2^{r_l}-1}} \qquad (4.33)$$

with P_b denoting the BER.

The available set of discrete rates and the SNR thresholds corresponding to BERs of 10^{-3} and 10^{-6} are shown in Tables 4.1 and 4.2, respectively.

The results of Fig. 4.3 show that the case of continuous rates outperforms the discrete rates scenario, as expected. For a given set of discrete rates, higher ergodic sum-rates can be achieved when the target BER increases, as can be seen in Fig. 4.3 by comparing the discrete rates case with BERs of 10^{-3} and 10^{-6} without channel coding.

TABLE 4.1 Discrete Rates and SNR Thresholds for a Target BER $= 10^{-3}$ and Four Discrete Rates

r_l (bits)	η_l (dB)
0	$-\infty$
2	9.97
4	16.96
6	23.19

TABLE 4.2 Discrete Rates and SNR Thresholds for a Target BER $= 10^{-6}$ and Four Discrete Rates

r_l (bits)	η_l (dB)
0	$-\infty$
2	13.60
4	20.59
6	26.82

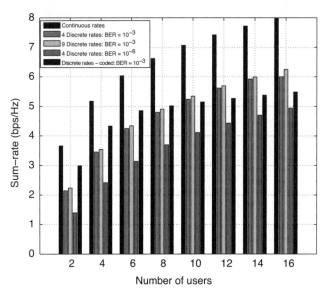

Figure 4.3 Comparison of optimal sum-rates with continuous and discrete rates.

4.4.3 Impact of Modulation and Coding Schemes

Increasing the number of available rates in the set of discrete rates by increasing the number of modulation schemes leads to the rates and SNR thresholds displayed in Table 4.3 for a BER of 10^{-3}. For a given target BER, increasing the number of discrete rates leads to results closer to optimal, as seen in Fig. 4.3 by comparing the four rates case to the nine rates case with a BER of 10^{-3}. However, the results show that this increase is marginal. This is due to the fact that the results presented are for uncoded MQAM modulation. Since the proposed solution framework only needs the set of discrete rates and the SNR thresholds, it can be easily applied to coded modulation.

The SNR thresholds of various coding schemes for the same number of modulation schemes used in Table 4.2 (QPSK, 16-QAM, and 64-QAM) obtained from

TABLE 4.3 Discrete Rates and SNR Thresholds for a Target BER $= 10^{-3}$ and Nine Discrete Rates

r_l (bits)	η_l (dB)
0	$-\infty$
1	5.2
2	9.97
3	13.65
4	16.96
5	20.11
6	23.19
7	26.24
8	29.27

TABLE 4.4   Discrete Rates and SNR Thresholds for a Target BER = 10^{-6} and 14 Modulation and Coding Schemes

MCS	r_l (bits)	η_l (dB)
No transmission	0	$-\infty$
QPSK, $R = 1/8$	0.25	-5.5
QPSK, $R = 1/5$	0.4	-3.5
QPSK, $R = 1/4$	0.5	-2.2
QPSK, $R = 1/3$	0.6667	-1.0
QPSK, $R = 1/2$	1.0	1.3
QPSK, $R = 2/3$	1.333	3.4
QPSK, $R = 4/5$	1.6	5.2
16-QAM, $R = 1/2$	2.0	7.0
16-QAM, $R = 2/3$	2.6667	10.5
16-QAM, $R = 4/5$	3.2	11.5
64-QAM, $R = 2/3$	4.0	14.0
64-QAM, $R = 3/4$	4.5	16.0
64-QAM, $R = 4/5$	4.8	17.0
64-QAM, $R = 1$ (uncoded)	6.0	26.8

Ref. [84] are shown in Table 4.4. Fig. 4.3 shows a comparison between the continuous rates case, the uncoded discrete rates cases for a BER of 10^{-6} (values obtained from Table 4.2), and the coded discrete rates case (values obtained from Table 4.4). Fig. 4.3 shows that the coded case achieves results closer to optimal than the uncoded case, but the performance gap between the coded and uncoded scenarios is reduced as the number of users increases. This is due to two main reasons:

1. The increase in the number of users leads to an increase in multiuser diversity that allows to achieve a better sum-rate with uncoded modulation.

2. The maximum achievable discrete rate is 6 bps/Hz (with uncoded 64-QAM) whereas there is no theoretical limit on the maximum rate achievable in the continuous rates scenario. This explains the decrease in the performance gap between the coded and uncoded discrete rates in addition to increase in the gap between the coded discrete rates case and the optimal continuous rates case.

It should be noted that in all the cases studied in Fig. 4.3, all users are allocated equal resources on average and achieve the same average rates, since they are located at the same distance from the BS. If this were not the case, most resources would be allocated to users closest to the BS since this would lead to sum-rate maximization, although at the expense of fairness. In this case, the weights π_k can be modified in order to give more priority to certain users. This is investigated in the following section for the two-users case.

4.4.4 Impact of Varying the User Weights

To illustrate the effect of weights on the resource allocation, a scenario of two users placed at equal distance from the BS is considered. The weight of user 1 is set to π_1 between 0 and 1, while the weight of user 2 is set to $\pi_2 = 1 - \pi_1$. The optimal solution derived in Section 4.3 is implemented with different user weights. The set of discrete rates $r_l \in \{0, 2, 4, 6\}$ bits with SNR thresholds $\eta_l \in \{-\infty, 9.97, 16.96, 23.19\}$ dB corresponding to a BER of 10^{-3} as in Table 4.1 is considered.

Fig. 4.4 shows the results of the simulation. As $\pi_1 < 0.5$, the rate of user 2 is greater than that of user 1. As expected, the break-even point is when the weights of both users are equal, that is, $\pi_1 = \pi_2 = 0.5$. In this case, the rates of user 1 and user 2 are equal since the weights are not playing any role in favoring one user over the other. Fig. 4.4 shows that each of the two users achieves a rate of 1.074 bps/Hz when the weights are equal. Furthermore, it should be noted that the maximum sum-rate is achieved when the weights are equal. Both users are allocated equal resources

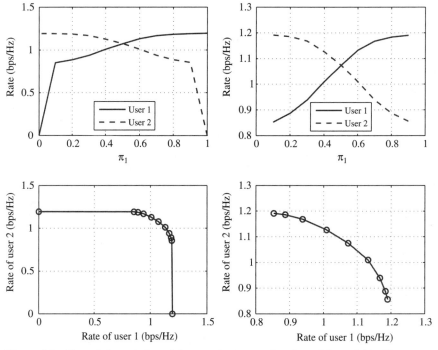

Figure 4.4 Upper left: evolution of the users' rates as a function of the weights. Upper right: evolution of the users' rates as a function of the weights without the cases $(\pi_1 = 1, \pi_2 = 0)$ and $(\pi_1 = 0, \pi_2 = 1)$. Lower left: rate of user 2 versus the rate of user 1. Lower right: rate of user 2 versus the rate of user 1 without the cases $(\pi_1 = 1, \pi_2 = 0)$ and $(\pi_1 = 0, \pi_2 = 1)$. The users are located at $d_1 = d_2 = 200$ m from the BS. This figure is reprinted, with permission from IEEE, from E. Yaacoub and Z. Dawy, Ergodic Sum-Rate Maximization in OFDMA Uplink with Discrete Rates, European Wireless Conference (EW 2010), Lucca, Italy, April 2010 ©2010 IEEE.

TABLE 4.5   Effect of Weights on the Users' Rates ($d_1 = 200$m, $d_2 = 300$m)

(π_1, π_2)	$(\lambda_1^*, \lambda_2^*)$	(R_1, R_2)(bps/Hz)
(0, 1)	(0.0001, 40.4578)	(0, 0.4730)
(0.1, 0.9)	(7.0903, 35.0784)	(1.0614, 0.4713)
(0.2, 0.8)	(14.3578, 29.9899)	(1.0827, 0.4692)
(0.3, 0.7)	(21.8718, 24.9417)	(1.1011, 0.4608)
(0.4, 0.6)	(29.8817, 20.3219)	(1.1319, 0.4450)
(0.5, 0.5)	(38.2717, 15.9561)	(1.1592, 0.4225)
(0.6, 0.4)	(47.0017, 12.2301)	(1.1769, 0.3990)
(0.7, 0.3)	(56.0923, 8.9104)	(1.1887, 0.3842)
(0.8, 0.2)	(65.0301, 5.7912)	(1.1935, 0.3743)
(0.9, 0.1)	(74.0057, 2.8674)	(1.1944, 0.3679)
(1, 0)	(83.0846, 0.0001)	(1.1951, 0)

This table is reprinted, with permission from IEEE, from E. Yaacoub and Z. Dawy, Ergodic Sum-Rate Maximization in OFDMA Uplink with Discrete Rates, European Wireless Conference (EW 2010), Lucca, Italy, April 2010 ©2010 IEEE.

on average and achieve the same average rates, since they are located at the same distance from the BS. If this were not the case, most resources would be allocated to the user closest to the BS since this would lead to sum-rate maximization, although at the expense of fairness. In this case, the weights π_k can be modified in order to give more priority to certain users.

To investigate the effect of the weights on the users' rates and therefore on the fairness in the system, the scenario of two users, with one placed at 200 m from the BS, and the other at 300 m from the BS, is considered. The weights of these users are varied between 0 and 1. The results are shown in Table 4.5. The weight of user 2, which is placed farther than user 1 from the BS, has allowed that user to be allocated some subcarriers and therefore achieve some rate particularly when its corresponding weight is high relative to that of user 1. However, as the weight of user 2 tends to 0, it provides less advantage to the user, since a low weight corresponds to imposing a penalty on user 2 that already has unfavorable channel conditions. When a user has a weight of 0, its rate does not have any contribution in the weighted maximization and hence the rate is set to zero in order to allocate all resources to the other user. Given that user 1 is closer to the BS than user 2, while a low weight decreases user 1's rate by almost 0.133 bps/Hz from the maximum achievable rate, its rate remains larger than that of user 2. These results show that giving higher weights to user 2 that is less favored by its channel conditions, allows it to enhance its performance compared to the equal weights scenario. This can be seen by comparing, in Table 4.5, the results with ($\pi_1 = 0.1$, $\pi_2 = 0.9$) to those with ($\pi_1 = 0.5$, $\pi_2 = 0.5$).

4.5   SUMMARY

The problem of weighted ergodic sum-rate maximization in the uplink of OFDMA systems was formulated and an analytical solution was derived. Conversely to the continuous rates scenario, the problem investigated consists of a limited achievable number of discrete rates. Numerical results showed that results became closer to the optimal continuous rates case when the target bit error rate increased or when the number of discrete rates increased. Furthermore, results showed that varying the weights allows providing more fairness in the system by increasing the priority of unfavored users, and that the maximum sum-rate is achieved when the weights are equal.

In the next chapter, the solutions of Chapters 3 and 4 are extended to a wide class of utility functions, where fairness can be embedded in the utility function itself. Chapter 5 constitutes an intermediate step between Chapters 3 and 4 on one hand, and Chapters 6 and 7 on the other hand, where low complexity utility maximization algorithms are presented, and where sum-rate maximization represents a special case of the utility function.

GENERALIZATION TO UTILITY MAXIMIZATION

Chapters 3 and 4 presented the optimal solutions for weighted ergodic sum-rate maximization with continuous and discrete rates, respectively. In this chapter, the solutions are extended to general utility maximization, where the utility of a user is a function of its rate. The utilities considered in this chapter are assumed to be concave or quasi-concave, and are assumed to be separable across subcarriers. In Chapter 6, suboptimal scheduling algorithms are presented for general utility maximization, where no particular assumptions or constraints are imposed on the utility function. In Chapter 6, the proposed algorithms are applied to the case of sum-rate maximization with equal user weights and no minimum rate constraints. These algorithms are used in Chapter 7 for proportional fair scheduling, where the utilities used are not separable among the subcarriers. Proportional fair scheduling allows to ensure more fairness between users in the resource allocation process. The chapter is organized as follows. Section 5.1 presents the necessary background and lists the main contributions of this chapter. Utility maximization for continuous rates is discussed in Section 5.2, whereas utility maximization for discrete rates is discussed in Section 5.3. The chapter is summarized in Section 5.4.

5.1 BACKGROUND

In Chapter 3, the optimal solution for weighted ergodic sum-rate maximization with continuous rates was presented, whereas the solution for the discrete rates case was presented in Chapter 4. In this chapter, these results are extended to general utility maximization, where the utility of a user is a function of its rate.

In this chapter, the sum-utility maximization is considered for a class of utility functions satisfying the following conditions: the utility is a concave increasing function of the rate, it is equal to zero when the rate is zero, and it is separable across

Resource Allocation in Uplink OFDMA Wireless Systems: Optimal Solutions and Practical Implementations,
Elias E. Yaacoub and Zaher Dawy.
© 2012 by the Institute of Electrical and Electronics Engineers, Inc. Published 2012 by John Wiley & Sons, Inc.

subcarriers, that is, we can write

$$U(\mathbf{A}, \mathbf{P}) = \sum_{k=1}^{K} U_k(R_k)$$

$$= \sum_{k=1}^{K} \sum_{i=1}^{N} \alpha_{k,i} U_{k,i} \left(R_{k,i}(P_{k,i}, g_{k,i}) \right)$$

(5.1)

The main topics discussed in this chapter are summarized as follows:

1. Extending sum-rate maximization with continuous rates to the case of utility maximization with continuous rates, and presenting the optimal solution for the class of concave separable utilities.

2. Extending sum-rate maximization with discrete rates to the case of utility maximization with discrete rates, and presenting the optimal solution for the class of concave or quasiconcave separable utilities.

5.2 ERGODIC UTILITY MAXIMIZATION WITH CONTINUOUS RATES

In Chapter 3, the solution of the ergodic sum-rate maximization problem with continuous rates was derived. In this section, this solution is extended to utility maximization, with the utility satisfying the conditions described in Section 5.1.

The utility maximization problem with continuous rates is expressed as

$$\max_{\mathbf{A}, \mathbf{P}} \mathbb{E}_g \left\{ \sum_{k=1}^{K} \sum_{i=1}^{N} \alpha_{k,i} U_{k,i} \left(R_{k,i}(P_{k,i}, g_{k,i}) \right) \right\}$$

(5.2)

subject to

$$\mathbb{E}_g \left\{ \sum_{i=1}^{N} P_{k,i} \right\} \leq P_k, \quad \text{for all } k$$

$$\sum_{k=1}^{K} \alpha_{k,i} \leq 1, \quad \text{for all } i$$

Sum-rate maximization is a special case of utility maximization, where the utility considered is equal to the rate, that is, $U_{k,i} \left(R_{k,i}(P_{k,i}, g_{k,i}) \right) = R_{k,i}(P_{k,i}, g_{k,i})$. Weighted sum-rate maximization considered in Chapter 3 is another special case, where the utility is the rate multiplied by a weighting factor to differentiate between the priorities of the different users, that is, $U_{k,i} \left(R_{k,i}(P_{k,i}, g_{k,i}) \right) = \pi_k R_{k,i}(P_{k,i}, g_{k,i})$. Minimum rate constraints are not considered in (5.2) since the purpose of such constraints is to ensure fairness in resource allocation. With utility maximization, fairness constraints are assumed to be incorporated within the utility function $U(\cdot)$ itself. In other words, the utility function can be selected in a way that guarantees a certain fairness in the allocation of resources. Examples of such utilities are given in Chapter 7.

The problem in (5.2) is nonconvex. Relaxing the condition on the $\alpha_{k,i}$s and allowing them to take any value in the interval [0, 1] instead of the set {0, 1} (which is equivalent to allowing time-sharing of a single subcarrier between different users), and letting $f_{k,i} = \alpha_{k,i} P_{k,i}$, the problem in (5.2) can be reformulated as follows:

$$\max_{\mathbf{A,F}} \mathbb{E}_g \left\{ \sum_{k=1}^{K} \sum_{i=1}^{N} \alpha_{k,i} U_{k,i} \left(\frac{f_{k,i}}{\alpha_{k,i}}, g_{k,i} \right) \right\} \tag{5.3}$$

subject to

$$\mathbb{E}_g \left\{ \sum_{i=1}^{N} f_{k,i} \right\} \le P_k, \quad \text{for all } k$$

$$\sum_{k=1}^{K} \alpha_{k,i} \le 1, \quad \text{for all } i$$

The problem in (5.3) is clearly convex since expectation preserves convexity and the concavity of U is preserved by the perspective operation $(\alpha U(f/\alpha))$ [34]. The problem in (5.3) is equivalent to the original problem in (5.2) when the condition on $\alpha_{k,i}$s is relaxed: for each user k and subcarrier i, finding the optimal pair $(\alpha_{k,i}^*, f_{k,i}^*) = (\alpha_{k,i}^*, \alpha_{k,i}^* P_{k,i}^*)$ leads to the same solution as finding $(\alpha_{k,i}^*, P_{k,i}^*)$.

In order to reach the solution of the problem defined in (5.3), the Lagrangian, which entails a vector of Lagrange multipliers λ corresponding to the power constraint, is defined.

$$L(\mathbf{A, F}, \lambda) = \mathbb{E}_g \left\{ \sum_{k=1}^{K} \sum_{i=1}^{N} \alpha_{k,i} U_{k,i} \left(R_{k,i}(f_{k,i}, g_{k,i}) \right) \right\}$$
$$- \sum_{k=1}^{K} \lambda_k \left(\mathbb{E}_g \left\{ \sum_{i=1}^{N} f_{k,i} \right\} - P_k \right) \tag{5.4}$$

The Lagrangian dual function is then given by

$$D(\lambda) = \max_{\mathbf{A,F}} L(\mathbf{A, F}, \lambda) \tag{5.5}$$

The optimization dual problem is given by

$$\min_{\lambda > 0} D(\lambda) \tag{5.6}$$

The Lagrangian dual can be written, after some straightforward mathematical manipulations, as

$$D(\lambda) = \max_{\mathbf{A}} \sum_{k=1}^{K} \sum_{i=1}^{N} \max_{\mathbf{P}} \mathbb{E}_g \left\{ \alpha_{k,i} \Phi(P_{k,i}) \right\} + \sum_{k=1}^{K} \lambda_k P_k \tag{5.7}$$

with $\Phi(P_{k,i})$ defined as follows:

$$\Phi(P_{k,i}) = U_{k,i}\left(R_{k,i}(P_{k,i}, g_{k,i})\right) - \lambda_k P_{k,i} \tag{5.8}$$

To maximize (5.7) for any given $\alpha_{k,i}$, (5.8) is differentiated with respect to $P_{k,i}$ and the result is set to 0. This yields

$$P^*_{k,i} = \left[\frac{U'_{k,i}(R_{k,i})}{\ln(2)\lambda_k} - \frac{1}{g_{k,i}}\right]^+ \tag{5.9}$$

where $[y]^+ = \max(0, y)$.

The solution in (5.9) corresponds to the maximum per each fading state realization and hence for the expectation over g [33]. This solution is a water-filling solution that depends on the multiplier λ_k associated with the per-user power constraint, the derivative $U'_{k,i}(R_{k,i})$ of the utility with respect to the rate of user k on subcarrier i, and the channel gain of user k on subcarrier i. In the case of sum-rate maximization, the derivative is given by $U'_{k,i}(R_{k,i}) = 1$, since $U_{k,i}(R_{k,i}) = R_{k,i}$. In such a scenario, (5.9) corresponds to the standard water-filling solution. In weighted ergodic sum-rate maximization with minimum rate constraints, the utility is expressed by $U_{k,i}(R_{k,i}) = \pi_k R_{k,i} + \mu_k(R_{k,i} - R_{k,\min})$ and thus $U'_{k,i}(R_{k,i}) = (\pi_k + \mu_k)$, which corresponds to the solution obtained in (3.13).

Following the approach of Section 3.3.1, the optimal subcarrier allocation is given by

$$\alpha^*_{k,i}(g) = \begin{cases} 1, & \text{if } k = k^*_i \text{ for } i = 1, ..., N \\ 0, & \text{otherwise} \end{cases} \tag{5.10}$$

where

$$k^*_i = \arg\max_k U_{k,i}\left(R_{k,i}(P^*_{k,i}, g_{k,i})\right) - \lambda_k P^*_{k,i} \tag{5.11}$$

Hence, the optimal power allocation is given by

$$f^*_{k,i} = \alpha^*_{k,i} P^*_{k,i} = \begin{cases} P^*_{k,i}, & \text{if } k = k^*_i \\ 0, & \text{otherwise} \end{cases} \tag{5.12}$$

To complete the solution, the vector of geometric multipliers associated with the power constraint should be determined. Determining the multipliers can be done in the same way as in Section 3.3.1.

5.2.1   Duality Gap

In this section, the solution derived previously is shown to be unique and that it corresponds to zero duality gap. Consequently, the solution of the dual problem can be directly mapped to the solution of the primal utility maximization problem.

Using the obtained power and subcarrier allocation, the dual solution can be formulated as follows:

$$d^* = \sum_{k=1}^{K}\sum_{i=1}^{N} \mathbb{E}_g\left\{\alpha_{k,i}^* U_{k,i}\left(R_{k,i}(P_{k,i}^*, g_{k,i})\right) - \lambda_k^* \alpha_{k,i}^* P_{k,i}^*\right\} + \sum_{k=1}^{K}\lambda_k^* P_k \quad (5.13)$$

The average power constraint states that the average power used by each user should be less than the user's target average power P_k. The average power of the user is denoted by

$$P_k^{\text{avg}}(\lambda_k^*) = \mathbb{E}_g\left\{\sum_{i=1}^{N}\alpha_{k,i}^* P_{k,i}^*\right\} \quad (5.14)$$

The dual solution can therefore be rewritten as

$$d^* = \sum_{k=1}^{K}\sum_{i=1}^{N} \mathbb{E}_g\left\{\alpha_{k,i}^* U_{k,i}\left(R_{k,i}(P_{k,i}^*, g_{k,i})\right)\right\}$$
$$+ \sum_{k=1}^{K}\lambda_k^*\left(P_k - P_k^{\text{avg}}(\lambda_k^*)\right) \quad (5.15)$$

The primal solution can be written as

$$p^* = \mathbb{E}_g\left\{\sum_{k=1}^{K}\sum_{i=1}^{N}\alpha_{k,i}^* U_{k,i}\left(R_{k,i}(P_{k,i}^*, g_{k,i})\right)\right\} \quad (5.16)$$

From the weak duality theorem [35], given a problem with a dual solution d^* and a primal solution p^*, then

$$p^* \leq d^* \quad (5.17)$$

The duality gap is given by

$$d^* - p^* = \sum_{k=1}^{K}\sum_{i=1}^{N} \mathbb{E}_g\left\{\alpha_{k,i}^* U_{k,i}\left(R_{k,i}(P_{k,i}^*, g_{k,i})\right)\right\}$$
$$+ \sum_{k=1}^{K}\lambda_k^*\left(P_k - P_k^{\text{avg}}(\lambda_k^*)\right)$$
$$- \mathbb{E}_g\left\{\sum_{k=1}^{K}\sum_{i=1}^{N}\alpha_{k,i}^* U_{k,i}\left(R_{k,i}(P_{k,i}^*, g_{k,i})\right)\right\} \quad (5.18)$$

From this equation, it can be easily shown that the duality gap is zero whenever

$$P_k^{\text{avg}} = P_k, \quad \text{for all } k \quad (5.19)$$

Thus, it can be asserted that if there exists a vector of λ_k's solution to the dual problem such that for each user k, $P_k^{\text{avg}} = P_k$, the duality gap between the primal and dual problems is 0. In fact, when λ_k is such that $P_k^{\text{avg}} = P_k$, the dual problem is minimized with respect to λ (see (5.4)–(5.6)). In addition, when $P_k^{\text{avg}} = P_k$, the primal problem is maximized, since the utility is an increasing function of the rate, and the rate (and hence the utility) is maximized when users transmit at the maximum allowed power. This is true because the subcarrier allocations are orthogonal at every fading state, and hence the transmission of one user does not interfere with the transmissions of other users.

5.3 ERGODIC UTILITY MAXIMIZATION WITH DISCRETE RATES

The problem of the maximization of the ergodic sum-utility with discrete rates can be formulated as follows:

$$\max_{\mathbf{A},\mathbf{P}} \mathbb{E}_g \left\{ \sum_{k=1}^{K} \sum_{i=1}^{N} \alpha_{k,i} U_{k,i} \left(R_{k,i}^d(P_{k,i}, g_{k,i}) \right) \right\} \tag{5.20}$$

subject to

$$\mathbb{E}_g \left\{ \sum_{i=1}^{N} P_{k,i} \right\} \leq P_k, \quad \text{for all } k$$

$$\sum_{k=1}^{K} \alpha_{k,i} \leq 1, \quad \text{for all } i$$

Ergodic sum-rate maximization is a special case of (5.20) where the utility considered is the rate itself. Different classes of users, where users in each class are given equal priority by the system, with different priorities between classes, for example [37], can be taken into account by considering $U_k = \pi_k R_k$, where the π_k's are priority weights fixed by higher layers.

Since the utility functions are concave and increasing with the rates, and since the rate function is quasiconcave, then the function $U_{k,i}(R_{k,i}^d)$ is quasiconcave since composition preserves quasi-concavity [34]. Instead of finding a family of concave functions $\phi_t(\eta_{k,i})$ then solving the problem via several iterations of convex feasibility problems as described in Ref. [34], a single problem is solved by considering the Lagrangian dual of (5.20) and showing that the solution of the dual problem corresponds to the primal solution.

To solve (5.20), the problem is first reformulated into an equivalent form that leads to the same solution but is easier to solve. By letting

$$f_{k,i} = \alpha_{k,i} P_{k,i} \tag{5.21}$$

It can be easily shown that

$$\alpha_{k,i} U_{k,i} \left(R_{k,i}^d(P_{k,i}, g_{k,i}) \right) = \alpha_{k,i} U_{k,i} \left(R_{k,i}^d(\alpha_{k,i} P_{k,i}, g_{k,i}) \right)$$
$$= \alpha_{k,i} U_{k,i} \left(R_{k,i}^d(f_{k,i}, g_{k,i}) \right) \tag{5.22}$$

since $\alpha_{k,i}$ can only take 0 or 1 values and $U_{k,i}(R_{k,i}^d) = 0$ if $R_{k,i}^d = 0$. Consequently, the problem in (5.20) is equivalent to

$$\max_{\mathbf{A},\mathbf{F}} \mathbb{E}_g \left\{ \sum_{k=1}^{K} \sum_{i=1}^{M} \alpha_{k,i} U_{k,i} \left(R_{k,i}^d(f_{k,i}, g_{k,i}) \right) \right\} \tag{5.23}$$

subject to

$$\mathbb{E}_g \left\{ \sum_{i=1}^{N} f_{k,i} \right\} \leq P_k, \quad \text{for all } k$$

$$\sum_{k=1}^{K} \alpha_{k,i} \leq 1, \quad \text{for all } i$$

since finding an optimal allocation $(\alpha_{k,i}^*, \alpha_{k,i}^* P_{k,i}^*)$ in (5.23) is equivalent to finding $(\alpha_{k,i}^*, P_{k,i}^*)$ in (5.20).

In order to reach the solution of the problem defined in (5.23), the Lagrangian entailing a vector of Lagrange multipliers λ corresponding to the power constraint is defined.

$$L(\mathbf{A}, \mathbf{F}, \lambda) = \mathbb{E}_g \left\{ \sum_{k=1}^{K} \sum_{i=1}^{N} \alpha_{k,i} U_{k,i} \left(R_{k,i}^d(f_{k,i}, g_{k,i}) \right) \right\}$$
$$- \sum_{k=1}^{K} \lambda_k \left(\mathbb{E}_g \left\{ \sum_{i=1}^{N} f_{k,i} \right\} - P_k \right) \tag{5.24}$$

The Lagrangian dual function is then given by

$$D(\lambda) = \max_{\mathbf{A},\mathbf{F}} L(\mathbf{A}, \mathbf{F}, \lambda) \tag{5.25}$$

The optimization dual problem is given by

$$\min_{\lambda > 0} D(\lambda) \tag{5.26}$$

The Lagrangian dual can be written, after some straightforward mathematical manipulations, as

$$D(\lambda) = \max_{\mathbf{A}} \sum_{k=1}^{K} \sum_{i=1}^{N} \max_{\mathbf{P}} \mathbb{E}_g \left\{ \alpha_{k,i} \Phi(P_{k,i}) \right\} + \sum_{k=1}^{K} \lambda_k P_k \tag{5.27}$$

with $\Phi(P_{k,i})$ defined as follows:

$$\Phi(P_{k,i}) = U_{k,i} \left(R_{k,i}^d(P_{k,i}, g_{k,i}) \right) - \lambda_k P_{k,i} \tag{5.28}$$

Consequently,

$$
\begin{aligned}
\Phi(P_{k,i}) &= U_{k,i}\left(R_{k,i}^d(P_{k,i}, g_{k,i})\right) - \lambda_k P_{k,i} \\
&= U_{k,i}(r_l) - \lambda_k P_{k,i} \\
&\leq U_{k,i}(r_l) - \lambda_k \frac{\eta_l}{g_{k,i}}
\end{aligned}
\tag{5.29}
$$

Due to the discrete nature of the rates, there are L candidate power allocation functions

$$
P_{k,i} \in \left\{ \frac{\eta_0}{g_{k,i}}, \frac{\eta_1}{g_{k,i}}, \ldots, \frac{\eta_{L-1}}{g_{k,i}} \right\}
\tag{5.30}
$$

from which the one maximizing (5.29) should be chosen. Hence,

$$
l_{k,i}^* = \arg\max_l U_{k,i}(r_l) - \lambda_k \frac{\eta_l}{g_{k,i}}
\tag{5.31}
$$

Then,

$$
R_{k,i}^* = r_{l_{k,i}^*}
\tag{5.32}
$$

and

$$
P_{k,i}^* = \frac{\eta_{l_{k,i}^*}}{g_{k,i}}
\tag{5.33}
$$

Due to the exclusivity of subcarrier allocation, (5.28) is maximized by allocating subcarrier i to user k_i^* such that

$$
\begin{aligned}
k_i^* &= \arg\max_k U_{k,i}(r_{l_{k,i}^*}) - \lambda_k P_{k,i}^* \\
&= \arg\max_k U_{k,i}(r_{l_{k,i}^*}) - \lambda_k \frac{\eta_{l_{k,i}^*}}{g_{k,i}}
\end{aligned}
\tag{5.34}
$$

Hence, the optimal subcarrier allocation is given by

$$
\alpha_{k,i}^* =
\begin{cases}
1, & \text{if } k = k_i^* \\
0, & \text{otherwise}
\end{cases}
\tag{5.35}
$$

and the optimal power allocation is given by

$$
f_{k,i}^* = \alpha_{k,i}^* P_{k,i}^* =
\begin{cases}
\dfrac{\eta_{l_{k,i}^*}}{g_{k,i}}, & \text{if } k = k_i^* \\
0, & \text{otherwise}
\end{cases}
\tag{5.36}
$$

In (5.31) and (5.34), λ_k acts as a power price that penalizes the user as the transmitted power increases. To complete the solution, the vector of geometric multipliers λ associated with the power constraint should be determined. Determining the multipliers can be done using an iterative subgradient method as in Section 4.3.

5.3.1 Duality Gap

In this section, the solution derived above for utility maximization with discrete rates is shown to be unique and corresponding to zero duality gap. Consequently, the solution of the dual problem can be directly mapped to the solution of the primal utility maximization problem.

Using the obtained power and subcarrier allocation, the dual solution can be formulated as follows:

$$d^* = \sum_{k=1}^{K} \sum_{i=1}^{N} \mathbb{E}_g \left\{ \alpha_{k,i}^* U_{k,i} \left(R_{k,i}^d(P_{k,i}^*, g_{k,i}) \right) - \lambda_k^* \alpha_{k,i}^* P_{k,i}^* \right\} + \sum_{k=1}^{K} \lambda_k^* P_k \quad (5.37)$$

The average power constraint states that the average power used by each user should be less than the user's target average power P_k. The average power of the user is denoted by

$$P_k^{avg}(\lambda_k^*) = \mathbb{E}_g \left\{ \sum_{i=1}^{N} \alpha_{k,i}^* P_{k,i}^* \right\} \quad (5.38)$$

The dual solution can therefore be rewritten as

$$d^* = \sum_{k=1}^{K} \sum_{i=1}^{N} \mathbb{E}_g \left\{ \alpha_{k,i}^* U_{k,i} \left(R_{k,i}^d(P_{k,i}^*, g_{k,i}) \right) \right\}$$
$$+ \sum_{k=1}^{K} \lambda_k^* \left(P_k - P_k^{avg}(\lambda_k^*) \right) \quad (5.39)$$

The primal solution can be written as

$$p^* = \mathbb{E}_g \left\{ \sum_{k=1}^{K} \sum_{i=1}^{N} \alpha_{k,i}^* U_{k,i} \left(R_{k,i}^d(P_{k,i}^*, g_{k,i}) \right) \right\} \quad (5.40)$$

From the weak duality theorem [35], given a problem with a dual solution d^* and a primal solution p^*, then

$$p^* \le d^* \quad (5.41)$$

The duality gap is given by

$$d^* - p^* = \sum_{k=1}^{K} \sum_{i=1}^{N} \mathbb{E}_g \left\{ \alpha_{k,i}^* U_{k,i} \left(R_{k,i}^d(P_{k,i}^*, g_{k,i}) \right) \right\}$$
$$+ \sum_{k=1}^{K} \lambda_k^* \left(P_k - P_k^{avg}(\lambda_k^*) \right) \quad (5.42)$$
$$- \mathbb{E}_g \left\{ \sum_{k=1}^{K} \sum_{i=1}^{N} \alpha_{k,i}^* U_{k,i} \left(R_{k,i}^d(P_{k,i}^*, g_{k,i}) \right) \right\}$$

From this equation, it can be easily shown that the duality gap is zero whenever

$$P_k^{\text{avg}} = P_k, \quad \text{for all } k \tag{5.43}$$

Thus, it can be asserted that if there exists a vector of λ_k's solution to the dual problem such that for each user k, $P_k^{\text{avg}} = P_k$, the duality gap between the primal and dual problems is 0. In fact, when λ_k is such that $P_k^{\text{avg}} = P_k$, the dual problem is minimized with respect to λ (see (5.24)–(5.26)). In addition, when $P_k^{\text{avg}} = P_k$, the primal problem is maximized, since the utility is an increasing function of the rate, and the rate (and hence the utility) is maximized when users transmit at the maximum allowed power. This is true because the subcarrier allocations are orthogonal at every fading state, and hence the transmission of one user does not interfere with the transmissions of other users. The solution in (5.34) is unique since we assume the fading distribution has a continuous density as in Section 4.3.1.

5.4   SUMMARY

In this chapter, the sum-rate maximization results of Chapters 3 and 4 were extended to the case of general utility maximization for both continuous and discrete rates. The utilities were assumed to be concave and separable across subcarriers. In the next chapter, Chapter 6, suboptimal scheduling algorithms are presented for general utility maximization, where no particular assumptions or constraints are imposed on the utility functions. The algorithms of Chapter 6 are used in Chapter 7 for proportional fair scheduling, where the utilities used are not separable among the subcarriers.

SUBOPTIMAL IMPLEMENTATION OF ERGODIC SUM-RATE MAXIMIZATION

Chapters 3 and 4 presented the optimal solutions for weighted ergodic sum-rate maximization with continuous and discrete rates, respectively. In Chapter 5, the derivations of Chapters 3 and 4 were extended to the case of utility maximization, where the utility functions satisfied the conditions listed in Section 5.1. In this chapter, suboptimal scheduling algorithms for utility maximization are presented. The algorithms do not assume any conditions on the utility functions considered. The results presented in this chapter correspond to the case of sum-rate maximization with equal user weights and no minimum rate constraints. The implementation of the proposed algorithms with utilities incorporating fairness will be discussed in Chapter 7. This chapter is organized as follows. Section 6.1 presents the necessary background and lists the main contributions of this chapter. The suboptimal scheduling algorithms for both the continuous and discrete rates scenarios are discussed in Sections 6.2 and 6.3, respectively. The complexity of the algorithms is analyzed in Section 6.4. Numerical results are presented and analyzed in Section 6.5. Finally, Section 6.6 summarizes the chapter.

6.1 BACKGROUND

In Chapter 3, the optimal solution for weighted ergodic sum-rate maximization with continuous rates was presented, whereas the solution for the discrete rates case was presented in Chapter 4. The derivations of Chapters 3 and 4 were extended to utility maximization in Chapter 5. These solutions require iterative subgradient iterations at the initiation phase, and a tracking of the channel probability density function to repeat the calculations when necessary. The computational load increases at the BS with the number of users, since the number of Lagrangian parameters increases with the number of users. Furthermore, the optimal solution necessitates that the transmit power be communicated to the users on each allocated subcarrier every TTI, since the optimal solution is determined by water-filling. This requires an increased signaling

Resource Allocation in Uplink OFDMA Wireless Systems: Optimal Solutions and Practical Implementations, Elias E. Yaacoub and Zaher Dawy.
© 2012 by the Institute of Electrical and Electronics Engineers, Inc. Published 2012 by John Wiley & Sons, Inc.

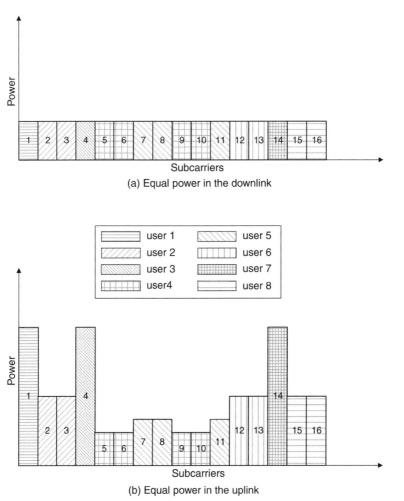

Figure 6.1 Comparison of equal power allocation in the downlink and in the uplink: resource allocation example with eight users and 16 subcarriers. The power per subcarrier in the downlink is P_{BS}/N and in the uplink $P_k/|\mathcal{I}_{\text{sub},k}|$, with P_{BS} the transmit power of the BS, P_k the transmit power of user k, N the number of subcarriers, $\mathcal{I}_{\text{sub},k}$ the set of subcarriers allocated to user k, and $|\cdot|$ denoting set cardinality.

overhead. It would be interesting if the optimal solution can be approximated by suboptimal algorithms based on equal power allocation. In this case, no subgradient iterations are required, and the signaling overhead consists of informing the users of their allocated subcarriers without the transmit power on each subcarrier, since with equal power allocation each user subdivides its transmit power equally over its allocated subcarriers. For the downlink, it was shown that equal power allocation approximates the optimal ergodic solution for the continuous rates scenario, but not for the discrete rates case [26]. For the uplink it was shown that equal power allocation approximates the optimal instantaneous water-filling solution for the continuous rates scenario [95, 110].

Therefore, in this chapter, suboptimal solutions for the uplink ergodic solution with continuous and discrete rates are presented. In fact, due to the distributed nature of the power constraint in the uplink, equal power per subcarrier is not equivalent to equal power in the downlink, since generally different users are allocated a different number of subcarriers, as shown in Fig. 6.1. Hence, equal power allocation in the uplink provides more potential to approximate the optimal solution than in the downlink. This will be the main focus of this chapter, which has the following objectives:

1. Presenting a suboptimal scheduling algorithm based on equal power allocation instead of optimal power allocation for the continuous rates scenario, and showing that the suboptimal algorithm performs comparably to the optimal solution for both instantaneous and ergodic sum-rate maximization.

2. Presenting suboptimal scheduling algorithms for the discrete rates scenario. The accuracy of the suboptimal algorithms in approximating the optimal solution is studied. Several scenarios are investigated by varying the number of users, the target BER, the number of modulation schemes, and the number of coding schemes.

3. Comparing the performance of the optimal discrete rates solution and of the suboptimal algorithms in the presence of imperfect CSI, and showing the robustness of the proposed suboptimal algorithms.

4. Comparing the performance of the suboptimal algorithms in the discrete rates scenario to other algorithms in the literature.

6.2   SUBOPTIMAL APPROXIMATION OF THE CONTINUOUS RATES SOLUTION

The proposed subcarrier allocation algorithms for both the continuous and discrete rates cases consist of allocating subcarrier i to user k in a way to maximize the difference

$$\Lambda_{k,i} = U(R_k|\mathcal{I}_{\text{sub},k} \cup \{i\}) - U(R_k|\mathcal{I}_{\text{sub},k}) \tag{6.1}$$

where the marginal utility, $\Lambda_{k,i}$, represents the gain in the utility function U when subcarrier i is allocated to user k, compared to the utility of user k before the allocation of i. The utility function can take different values depending on the scenario considered. In this chapter, the utility will correspond to the user rate. In Chapter 7, utilities corresponding to proportional fair scheduling will be investigated.

Suboptimal resource allocation in the continuous rates case is divided into two steps: subcarrier allocation to users followed by power allocation on the allocated subcarriers. Algorithm 6.1 describes the subcarrier allocation process in the continuous rates case.

Algorithm 6.1

> **Subcarrier Allocation Algorithm**
> **for** k **such that** $1 \le k \le K$
> $\quad \mathcal{I}_{\mathrm{sub},k} \leftarrow \emptyset$
> **end for**
> **for** i **such that** $1 \le i \le N$
> $\quad$ **for all** k **such that** $1 \le k \le K$
> $$U_k(P_k, \mathcal{I}_{\mathrm{sub},k} \cup \{i\}) \leftarrow \sum_{j \in (\mathcal{I}_{\mathrm{sub},k} \cup \{i\})} \log\left(1 + \frac{P_k}{|\mathcal{I}_{\mathrm{sub},k}| + 1} g_{k,j}\right)$$
> $\quad\quad$ **if** $\mathcal{I}_{\mathrm{sub},k} \neq \emptyset$
> $$U_k(P_k, \mathcal{I}_{\mathrm{sub},k}) \leftarrow \sum_{j \in \mathcal{I}_{\mathrm{sub},k}} \log\left(1 + \frac{P_k}{|\mathcal{I}_{\mathrm{sub},k}|} g_{k,j}\right)$$
> $\quad\quad$ **else**
> $\quad\quad\quad U_k(P_k, \mathcal{I}_{\mathrm{sub},k}) \leftarrow 0$
> $\quad\quad$ **end if**
> $\quad\quad \Lambda_{k,i} \leftarrow U_k(P_k, \mathcal{I}_{\mathrm{sub},k} \cup \{i\}) - U_k(P_k, \mathcal{I}_{\mathrm{sub},k})$
> $\quad$ **end for**
> $\quad k^* \leftarrow \arg\max_k \Lambda_{k,i}$
> $\quad \mathcal{I}_{\mathrm{sub},k^*} \leftarrow \mathcal{I}_{\mathrm{sub},k^*} \cup \{i\}$
> **end for**

Algorithm 6.2

> **Power Allocation Algorithm**
> **for** k **such that** $1 \le k \le K$
> $\quad$ **if** $\mathcal{I}_{\mathrm{sub},k} \neq \emptyset$
> $\quad\quad$ **for all** i **such that** $i \in \mathcal{I}_{\mathrm{sub},k}$
> $\quad\quad\quad P_{k,i} \leftarrow P_k / |\mathcal{I}_{\mathrm{sub},k}|$
> $\quad\quad$ **end for**
> $\quad$ **else**
> $\quad\quad$ **for all** i **such that** $1 \le i \le N$
> $\quad\quad\quad P_{k,i} \leftarrow 0$
> $\quad\quad$ **end for**
> $\quad$ **end if**
> **end for**

Algorithm 6.1 finds the user that has the highest marginal utility defined in (6.1) among all available users when the first available subcarrier is allocated to it. The process is repeated for all subcarriers. Then, the users allocate the power equally over the subcarriers assigned to them by Algorithm 6.1. The power allocation process is described by Algorithm 6.2.

Hence, the suboptimal resource allocation process for the continuous rates case is summarized in Algorithm 6.3.

Algorithm 6.3

> **Suboptimal Equal Power Scheduling with Continuous Rates**
> **Apply Algorithm   6.1 for subcarrier allocation**
> **Apply Algorithm   6.2 for power allocation**

Subdividing the power equally over the subcarriers is justified in Ref. [111] by the fact that the achieved gains are negligible compared to the increase in complexity when optimal power allocation is performed. In addition, it was shown in Refs [95, 110] that optimal power allocation using water-filling and equal power allocation over subcarriers lead to approximately the same results. In fact, in the uplink scenario, the maximum transmission power of mobile users is limited, contrarily to the downlink case where the BS has considerably more power and where the variation in user distances from the BS allows the latter to achieve gains by optimizing the power allocation.

6.3 SUBOPTIMAL APPROXIMATION OF THE DISCRETE RATES SOLUTION

Extending Algorithm 6.3 to the discrete rates case is not straightforward. Equal power allocation in the discrete rates scenario is more critical than in the continuous rates case: if, during the allocation process, enough power is available for user k to achieve a rate r_l with equal power allocation, then, at the end of the scheduling process, if user k is allocated more subcarriers, subdividing the power equally might not be sufficient to achieve r_l, and the achievable rates must be derived again. Hence, to reduce the complexity of the process, an initial subcarrier allocation based on continuous rates is performed. However, the utility used in the initial allocation constitutes an approximation of the discrete rates scenario for a given BER. In fact, for the rate calculations, the following expression is considered:

$$R(P_k, \mathcal{I}_{\text{sub},k}) = \sum_{i=1}^{N} \alpha_{k,i} \log_2(1 + \beta P_{k,i} g_{k,i}) \tag{6.2}$$

where $\alpha_{k,i} = 1$ if subcarrier i is allocated to user k, and β is called the SNR gap. It indicates the difference between the SNR needed to achieve a certain data transmission rate for a practical M-QAM system and the theoretical limit (Shannon capacity). It is given by [112]

$$\beta = \frac{-1.5}{\ln(5 P_b)} \tag{6.3}$$

where P_b denotes the BER. In (6.2), each user is assumed to transmit at the maximum power ($P_k = P_{k,\max}$), and the power is assumed to be subdivided equally among all the subcarriers allocated to that user. Hence, the SNR over a single subcarrier, $P_{k,i}g_{k,i}$, is given by

$$P_{k,i}g_{k,i} = \frac{\dfrac{P_k}{|\mathcal{I}_{\text{sub},k}|}H_{k,i}}{\sigma_i^2} \tag{6.4}$$

The initial allocation is implemented according to Algorithm 6.4. The difference in the subcarrier allocation method used in Algorithm 6.4 and that of Algorithm 6.1 is the use of (6.2) as a utility function instead of the ideal Shannon capacity formula corresponding to (6.2) with $\beta = 1$.

Algorithm 6.4

> **Continuous Rate Approximation of Discrete Rate Scheduling**
> **SNR Gap Computation:**
> $$\beta \leftarrow \frac{-1.5}{\ln(5\,P_b)}$$
> **Subcarrier Allocation:**
> **for** k **such that** $1 \le k \le K$
> $\mathcal{I}_{\text{sub},k} \leftarrow \emptyset$
> **end for**
> **for** i **such that** $1 \le i \le N$
>   **for all** k **such that** $1 \le k \le K$
>
> $$U_k(P_k, \mathcal{I}_{\text{sub},k} \cup \{i\}) \leftarrow \sum_{j \in (\mathcal{I}_{\text{sub},k} \cup \{i\})} \log\left(1 + \frac{\beta P_k}{|\mathcal{I}_{\text{sub},k}| + 1}g_{k,j}\right)$$
>
>   **if** $\mathcal{I}_{\text{sub},k} \ne \emptyset$
>
> $$U_k(P_k, \mathcal{I}_{\text{sub},k}) \leftarrow \sum_{j \in \mathcal{I}_{\text{sub},k}} \log\left(1 + \frac{\beta P_k}{|\mathcal{I}_{\text{sub},k}|}g_{k,j}\right)$$
>
>   **else**
> $U_k(P_k, \mathcal{I}_{\text{sub},k}) \leftarrow 0$
>   **end if**
> $\Lambda_{k,i} \leftarrow U_k(P_k, \mathcal{I}_{\text{sub},k} \cup \{i\}) - U_k(P_k, \mathcal{I}_{\text{sub},k})$
>   **end for**
> $k^* \leftarrow \arg\max_k \Lambda_{k,i}$
> $\mathcal{I}_{\text{sub},k^*} \leftarrow \mathcal{I}_{\text{sub},k^*} \cup \{i\}$
> **end for**
> **Power Allocation:**
> **Apply Algorithm   6.2 for power allocation**

After the subcarrier and power allocation according to Algorithm 6.4, the achievable discrete rate of every user on each subcarrier allocated to it can now be determined. This is a simple table lookup operation, where the rate of each user is $r_{l_{k,i}^*}$ such that $\eta_{l_{k,i}^*}$ is the highest SNR threshold below the actual SNR $P_{k,i}g_{k,i}$, with

$P_{k,i} = P_k / |\mathcal{I}_{\text{sub},k}|$, according to the subcarrier and power allocation of Algorithm 6.4. These steps are detailed in Algorithm 6.5.

Algorithm 6.5

> **Suboptimal Scheduling with Discrete Rates - 1**
> **Apply Algorithm   6.4 for subcarrier and power allocation**
> **for all** k **such that** $1 \le k \le K$
>> **for all** i **such that** $1 \le i \le N$
>>> $l_{k,i}^* \leftarrow \text{argmax}_{l, \eta_l \le P_{k,i} g_{k,i}} \eta_l$
>>> $R_{k,i}^d \leftarrow r_{l_{k,i}^*}$
>> **end for**
> **end for**

Algorithm 6.6

> **Suboptimal Scheduling with Discrete Rates - 2**
> **Apply Algorithm   6.5 for subcarrier and power allocation**
> **for all** k **such that** $1 \le k \le K$
>> $P_{\text{Buffer},k} \leftarrow 0$
>> **for all** i **such that** $i \in \mathcal{I}_{\text{sub},k}$
>>> $P_{\text{Buffer},k} \leftarrow P_{\text{Buffer},k} + (P_{k,i} - \eta_{l_{k,i}^*}/g_{k,i})$
>> **end for**
>> $STOP \leftarrow 0$ {**Variable to test the stopping criterion of the While loop**}
>> **while** $P_{\text{Buffer},k} > 0$ **AND** $STOP < |\mathcal{I}_{\text{sub},k}|$
>>> **for all** i **such that** $i \in \mathcal{I}_{\text{sub},k}$
>>>> **if** $(P_{\text{Buffer},k} - (\eta_{(l_{k,i}^*+1)}/g_{k,i} - \eta_{l_{k,i}^*}/g_{k,i})) > 0$
>>>>> $P_{\text{Buffer},k} \leftarrow P_{\text{Buffer},k} - (\eta_{(l_{k,i}^*+1)}/g_{k,i} - \eta_{l_{k,i}^*}/g_{k,i})$
>>>>> $l_{k,i}^* \leftarrow l_{k,i}^* + 1$
>>>>> $R_{k,i}^d \leftarrow r_{l_{k,i}^*}$
>>>>> $STOP \leftarrow 0$
>>>> **else**
>>>>> $STOP \leftarrow STOP + 1$
>>>> **end if**
>>> **end for**
>> **end while**
>> **if** $P_{\text{Buffer},k} > 0$
>>> **Distribute the remaining power equally on all allocated subcarriers:**
>>>> **for all** i **such that** $i \in \mathcal{I}_{\text{sub},k}$
>>>>> $P_{k,i} \leftarrow P_{k,i} + P_{\text{Buffer},k}/|\mathcal{I}_{\text{sub},k}|$
>>>> **end for**
>>> **end if**
> **end for**

The use of equal power with discrete rates according to Algorithm 6.5 leads to allocating, on each subcarrier, excessive power beyond the power necessary to achieve the rate $r_{l_{k,i}^*}$. In order not to "waste" this power, Algorithm 6.6 is proposed. Algorithm 6.5 has a reduced complexity over Algorithm 6.6, but the latter makes better use of the available power. Their performance will be compared in Section 6.5.

In Algorithm 6.6, for each user, the excessive power on each allocated subcarrier using equal power allocation is gathered into a sort of power "buffer". Then each user allocates the power of the buffer sequentially to its allocated subcarriers, whenever there is enough power in the buffer to increase the achievable rate into the next level. The process is repeated until no improvement can be made. After this process, the power remaining in the buffer is added in equal shares on all subcarriers, so that each user transmits at the maximum power $P_k = P_{k,\max}$. The benefits of this approach will be investigated in Section 6.5.4. It should be noted that the final outcome of Algorithm 6.6 does not necessarily lead to an equal power allocation, although it starts from an initial equal power allocation.

6.4 COMPLEXITY ANALYSIS OF THE SUBOPTIMAL ALGORITHMS

In this section, the complexity of the proposed scheduling algorithms is derived in the continuous and discrete rates scenarios, and the results are compared to the optimal solutions.

6.4.1 Complexity Analysis in the Continuous Rates Case

Algorithm 6.3 allocates each subcarrier after performing a linear search on the users in order to find the user that maximizes the marginal utility. Consequently, the total complexity of the algorithm is $\mathcal{O}(NK)$, that is, the algorithm has linear complexity in the number of users and in the number of subcarriers, and thus can be easily implemented in real-time.

On the other hand, the optimal solution derived in Chapter 3 requires a complexity of $\mathcal{O}(INK)$ in the Lagrangian parameters computation phase, with I the number of iterations. Afterward, the complexity becomes $\mathcal{O}(NK)$, similarly to the proposed algorithm. However, the proposed suboptimal algorithm is simpler to implement, since it does not involve water-filling computations, but rather the subdivision of the power equally among the allocated subcarriers. Furthermore, it does not require any Lagrangian computation, and thus does not need to keep track of the channel statistics in order to initiate a new Lagrangian computation whenever the channel probability density function changes. Thus, the proposed suboptimal algorithm is easier to be implemented in real-time than the optimal solution. Nevertheless, it should be noted that the optimal ergodic sum-rate maximization solution is still simpler to implement than the optimal instantaneous sum-rate maximization, which corresponds to (3.6) without the expectation. In fact, instantaneous sum-rate maximization requires a complexity of $\mathcal{O}(INK)$ at each scheduling interval, since the Lagrangian parameters need to be computed at every scheduling instant in order to obtain the optimal performance.

6.4.2 Complexity Analysis in the Discrete Rates Case

Algorithm 6.5 allocates each subcarrier after performing a linear search on the users. Then, it performs a table lookup operation in order to determine the achievable discrete rate. Consequently, the total complexity of the algorithm is $\mathcal{O}(NK \ln(L))$, that is, the algorithm has linear complexity in the number of users and in the number of subcarriers, in addition to a logarithmic complexity in the number of rates. Hence, it can be implemented in real-time.

On the other hand, the optimal solution derived in Chapter 4 requires a complexity of $\mathcal{O}(INKL)$ in the Lagrangian parameters computation phase, with I the number of iterations. Afterward, the complexity becomes $\mathcal{O}(NKL)$. However, it can be easily shown that (4.15) can be transformed into a table lookup operation, as proven in Ref. [26] for the downlink scenario. In this case, the complexity of the optimal solution at runtime becomes $\mathcal{O}(NK \ln(L))$, similarly to the Algorithm 6.5. However, the proposed suboptimal algorithm is simpler to implement, since it does not require any Lagrangian computation, and thus does not need to keep track of the channel statistics in order to initiate a new Lagrangian computation whenever the channel pdf changes. Thus, the proposed suboptimal algorithm is easier to be implemented in real-time than the optimal solution. Nevertheless, it should be noted that the optimal ergodic sum-rate maximization solution is still simpler to implement than the optimal instantaneous sum-rate maximization, which corresponds to (4.4) without the expectation. In fact, instantaneous sum-rate maximization requires a complexity of $\mathcal{O}(INKL)$ at each scheduling interval, since the Lagrangian parameters need to be computed at every scheduling instant in order to obtain the optimal performance.

The complexity of Algorithm 6.6 can be expressed as $\mathcal{O}(I_2 NK \ln(L))$, where I_2 is the number of iterations where the "while loop" of Algorithm 6.6 is implemented. In practice, I_2 is usually small, and during the generation of the results of Section 6.5, one or two iterations were always sufficient.

The complexity analysis of the optimal solutions and of the proposed algorithms is summarized in Table 6.1.

TABLE 6.1   Resource Allocation Complexity of the Optimal Solutions and the Suboptimal Algorithms

Algorithm	Initialization	Runtime
Continuous ergodic rates	$\mathcal{O}(INK)$	$\mathcal{O}(NK)$
Continuous instantaneous rates	–	$\mathcal{O}(INK)$
Continuous suboptimal (Algorithm 6.3)	–	$\mathcal{O}(NK)$
Discrete ergodic rates	$\mathcal{O}(INKL)$	$\mathcal{O}(NK \ln(L))$
Discrete instantaneous rates	–	$\mathcal{O}(INKL)$
Discrete suboptimal 1 (Algorithm 6.5)	–	$\mathcal{O}(NK \ln(L))$
Discrete suboptimal 2 (Algorithm 6.6)	–	$\mathcal{O}(I_2 NK \ln(L))$

6.5 RESULTS AND DISCUSSION

In this section, numerical results obtained by comparing the suboptimal algorithms presented in this chapter to the theoretical solutions of Chapters 3 and 4 are presented.

6.5.1 Simulation Parameters

For the simulations presented in this section, the fading is considered to have a Rayleigh distribution and hence the channel gain of each user is exponentially distributed with mean equal to the path gain representing propagation loss and given by $\kappa d_k^{-\upsilon}$, with κ a constant chosen to be -128.1 dB, d_k the distance in km from user k to the BS, and υ the path loss exponent, set to a value of 3.76. The scenario of users located on an equal distance, fixed to 200 m from the BS, is studied. The simulation model consists of a single BS having an omnidirectional antenna. The rate is averaged over 10,000 TTIs, with the duration of a TTI being 1 ms. The total bandwidth considered is $B = 5$ MHz, subdivided into $N = 16$ subcarriers. The maximum user transmit power is considered to be $P_k = 125$ mW for all k.

6.5.2 Results of the Continuous Rates Approximation

The optimal solution derived in Chapter 3 with an average power of 125 mW is compared to three different scenarios: the case of low complexity scheduling using Algorithm 6.3 and subdividing the total user transmit power equally among the subcarriers allocated to that user, the case of instantaneous sum-rate maximization corresponding to (3.7) without the expectation, which is equivalent to instantaneous sum-rate maximization under a peak power constraint (and hence corresponds to the solution derived in Ref. [97]), and the case of ergodic sum-rate subjected to an additional peak power constraint, such that the same instantaneous power is not allowed to exceed the maximum at each TTI. It should be noted that the power in ergodic sum-rate maximization could exceed the peak, since the constraint is on average power. Adding a peak power constraint to the problem is straightforward. However, due to the instantaneous nature of the peak power constraint, the expectation cannot be used, and hence the Lagrange multipliers corresponding to the peak constraint will have to be computed via subgradient iterations at each TTI (not offline), which hinders the idea of ergodicity and long-term average. For this reason, a suboptimal approach is used: it consists of scaling the power obtained by optimal ergodic sum-rate maximization to force the peak constraint without computing Lagrange multipliers at each TTI. On the other hand, for the case of instantaneous sum-rate maximization, the Lagrangian parameters λ_ks should be computed at every TTI, which leads to a considerable increase in complexity. In the simulations, the number of iterations is limited to a maximum of 500. In case convergence is not achieved when the maximum number of iterations is reached, the values of the λ_ks at the last iteration are used.

The results are shown in Fig. 6.2. For a small number of users (up to six), the case of ergodic sum-rate achieves the best results, since its instantaneous peak power can occasionally exceed the maximum transmit power, followed by the instantaneous

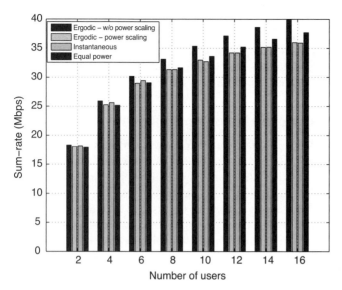

Figure 6.2 Sum-rate achieved by the different algorithms.

sum-rate maximization, since in this case the Lagrange multipliers are computed in each TTI and the values are optimized to each fading realization. The case of ergodic sum-rate with scaled maximum power comes third, since in this case the Lagrange multiplier of each user is computed offline and used through all the TTIs, which leads, on average, to a performance very close to that of the instantaneous case. As expected, the case of equal power distribution comes last.

However, it should be noted that all four scenarios have a very comparable performance. As the number of users increases, the superiority of ergodic sum-rate is maintained, but the equal power case outperforms the instantaneous case and the ergodic case with power scaling. This due to the fact that in the instantaneous case, when the number of users increase, the convergence to the optimal values of the λ_ks does not occur within the maximum number of iterations (which is set to 500 iterations per TTI). This leads to a suboptimal performance, although not far from the optimal case. For the ergodic case with power scaling, the increase in the number of users increases the probability that at a certain TTI, one or more users are not allocated any subcarriers, and hence their transmit power is zero. Although at other TTIs they are transmitting at $P_{k,\max}$, the long-term average power will be less than 125 mW, which scales down the performance.

Hence, Fig. 6.2 shows that ergodic sum-rate maximization achieves close results to instantaneous sum-rate maximization, at the expense of reduced complexity. In fact, when the number of users increases, the convergence of the instantaneous sum-rate maximization requires an increasing number of iterations, since a larger number of Lagrangian parameters λ_ks should be computed, due to the distributive nature of the power constraint in the uplink. Consequently, the case of ergodic sum-rate with power scaling to meet the peak power constraint could outperform the instantaneous sum-rate case when convergence is not reached, since with ergodic sum-rate the

λ_ks, computed once, can be used without recomputation as long as the channel pdf remains the same. Furthermore, Fig. 6.2 shows that the linear complexity suboptimal algorithm with equal power allocation can achieve a performance remarkably close to optimal, at a considerably reduced complexity. In addition, it can outperform the instantaneous sum-rate case when convergence is not reached.

6.5.3 Results of the Discrete Rates Approximation

In this section, the optimal discrete rates solution is compared to the suboptimal equal power algorithms presented in Sections 6.2 and 6.3. Algorithm 6.3 is used to approximate the continuous rates case and Algorithm 6.5 is used to approximate the discrete rates case. Fig. 6.3 shows the results of the optimal continuous rates solution and its equal power approximation using Algorithm 6.3, the discrete rates optimal solution corresponding to Table 4.1 and its equal power approximation using Algorithm 6.5, in addition to the discrete rates optimal solution corresponding to Table 4.2 and its equal power approximation using Algorithm 6.5. Fig. 6.3 shows that the equal power case achieves a performance close to optimal in the continuous rates scenario. However, there is an approximately constant performance gap of around 1 bps/Hz between the optimal discrete case and its equal power approximation. The relatively large gap is due to the limited number of discrete rates used, which leads to an increase of the power wasted using equal power allocation. Fig. 6.4 shows a comparison of the discrete rates of Tables 4.1 and 4.3 and their equal power approximations. Clearly, as the number of discrete rates increases, the optimal solution can be better approximated by the equal power algorithm. The results of Algorithm 6.4, which approximates the optimal discrete rates solution using continuous rates according to (6.2), are shown for comparison. Although equal power is used in Algorithm 6.4, the results

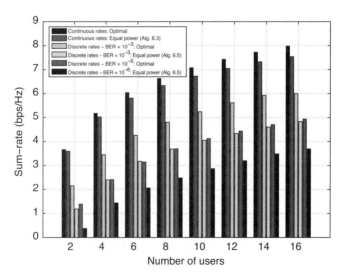

Figure 6.3 Comparison of sum-rates with continuous and discrete rates: optimal versus equal power allocation.

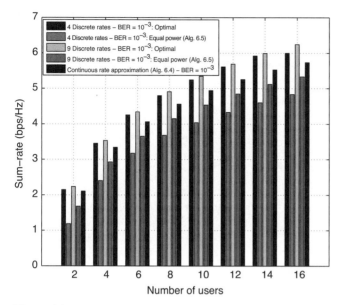

Figure 6.4 Comparison of sum-rates with discrete rates and a target BER $= 10^{-3}$: optimal versus equal power allocation.

of Fig. 6.4 show that it can approximate well the optimal discrete rates solution with optimal power allocation.

6.5.4 Results in the Case of Imperfect CSI

The results of Section 6.5.3 show that the optimal solution with discrete rates outperforms the equal power resource allocation with discrete rates. When the number of modulation schemes is reduced, the performance gap is shown to be relatively large. However, in Section 6.5.3, perfect CSI is assumed. In this section, the impact of imperfect or erroneous CSI on the scheduling process is investigated. Instead of the optimal value $g_{k,i}$, an erroneous estimate $\hat{g}_{k,i}$ is assumed to be available at the BS. The error $e_{k,i}$ in estimating $g_{k,i}$ is modeled by a Gaussian random variable with zero-mean and variance σ_e^2. The Assumption of Gaussian estimation error was also adopted in Refs [113–115]. This model is justified by the fact that the BS should be able to make measurements that are more likely to be close to the actual value of the CSI and less likely to be too far from the actual value. It is also justified by the fact that the error in many well-designed estimators is asymptotically Gaussian [115].

6.5.4.1 *Probability of Achieving a Reduced Rate*   With optimal resource allocation in the discrete rates scenario, underestimating the CSI will lead to assuming that a user k can achieve a rate r_l on subcarrier i and allocating the power accordingly. However, the power allocated might be sufficient to achieve a rate $r_{l'} \geq r_l$. But since the user is informed by the BS to use the modulation and coding scheme corresponding to r_l, only r_l will be achieved. On the other hand, overestimating the CSI will lead

to assuming that a user k can achieve a rate r_l on subcarrier i and allocating the power accordingly. However, the power allocated will be sufficient to achieve a rate $r_{l'} < r_l$.

With $\hat{g}_{k,i} = g_{k,i} + e_{k,i}$, and considering that user k assumes $r_{l_{k,i}^*}$ is achievable on subcarrier i, then, in the optimal case, the following scenarios are available:

- If $\hat{g}_{k,i} \leq g_{k,i}$ then $\left(P_{k,i} g_{k,i} = \dfrac{\eta_{l_{k,i}^*}}{\hat{g}_{k,i}} \cdot g_{k,i} \right) \geq \eta_{l_{k,i}^*}$, and $r_{l_{k,i}^*}$ is achievable.

- If $\hat{g}_{k,i} > g_{k,i}$ then $\left(P_{k,i} g_{k,i} = \dfrac{\eta_{l_{k,i}^*}}{\hat{g}_{k,i}} \cdot g_{k,i} \right) < \eta_{l_{k,i}^*}$, and $r_{l_{k,i}^*}$ is not achievable.

The probability of achieving a rate less than $r_{l_{k,i}^*}$ is given by

$$
\begin{aligned}
\text{Prob}\left(P_{k,i} g_{k,i} < \eta_{l_{k,i}^*} \right) &= \text{Prob}\left(\dfrac{\eta_{l_{k,i}^*}}{\hat{g}_{k,i}} \cdot g_{k,i} < \eta_{l_{k,i}^*} \right) \\
&= \text{Prob}\left(\hat{g}_{k,i} > g_{k,i} \right) \\
&= \text{Prob}\left(g_{k,i} + e_{k,i} > g_{k,i} \right) \\
&= \text{Prob}\left(e_{k,i} > 0 \right) \\
&= 1/2
\end{aligned}
\tag{6.5}
$$

where the last equality is due to the Gaussian distribution of $e_{k,i}$.

In the equal power case, the power transmitted by user k on subcarrier i is greater than $\dfrac{\eta_{l_{k,i}^*}}{\hat{g}_{k,i}}$. Hence, even when $\hat{g}_{k,i} > g_{k,i}$, $r_{l_{k,i}^*}$ might still be achievable. In this case, the probability of achieving a rate less than $r_{l_{k,i}^*}$ is given by

$$
\begin{aligned}
\text{Prob}\left(P_{k,i} g_{k,i} < \eta_{l_{k,i}^*} \right) &= \text{Prob}\left(P_{k,i}(\hat{g}_{k,i} - e_{k,i}) < \eta_{l_{k,i}^*} \right) \\
&= \text{Prob}\left(P_{k,i}\hat{g}_{k,i} - P_{k,i} e_{k,i} < \eta_{l_{k,i}^*} \right) \\
&= \text{Prob}\left(P_{k,i} e_{k,i} > P_{k,i}\hat{g}_{k,i} - \eta_{l_{k,i}^*} \right) \\
&= \text{Prob}\left(e_{k,i} > \hat{g}_{k,i} - \dfrac{\eta_{l_{k,i}^*}}{P_{k,i}} \right) \\
&= \int_{\hat{g}_{k,i} - \frac{\eta_{l_{k,i}^*}}{P_{k,i}}}^{\infty} \dfrac{1}{\sqrt{2\pi}\sigma_e} e^{-y^2/2\sigma_e^2} \, dy \\
&\leq 1/2
\end{aligned}
\tag{6.6}
$$

where the last inequality is obtained because $\hat{g}_{k,i} - \dfrac{\eta_{l_{k,i}^*}}{P_{k,i}} \geq 0$ since $P_{k,i} \geq \dfrac{\eta_{l_{k,i}^*}}{\hat{g}_{k,i}}$. Consequently, the suboptimal approach based on equal power allocation has a lower probability of achieving a reduced rate in the imperfect CSI scenario.

6.5.4.2 Simulation Results with Imperfect CSI

The performance of the optimal allocation schemes with discrete rates is compared to the suboptimal algorithms. The results are shown in Fig. 6.5. For the suboptimal scenario, in addition to Algorithm 6.5, which applies equal power allocation, the results of Algorithm 6.6 are

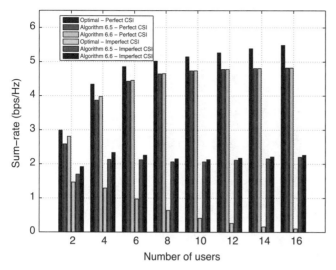

Figure 6.5   Comparison of discrete rates with channel coding and a target BER $= 10^{-6}$ in the case of perfect and imperfect CSI with $\sigma_e = 1$ dB: optimal versus suboptimal scheduling.

also shown. The standard deviation of the CSI error considered in the simulations is assumed to be 1 dB relative to the actual CSI. Fig. 6.5 compares the results of optimal and suboptimal scheduling using the rates of Table 4.4.

The results show that the suboptimal algorithms based on equal power allocation are robust to channel estimation errors, whereas the performance of the optimal solution deteriorates dramatically. When equal power allocation is used, the extra power that was assumed "wasted" in the perfect CSI case becomes a sort of power margin in the imperfect CSI case. This additional margin allows to cope (partially) with the channel estimation errors and leads to a considerably better performance for the suboptimal algorithms compared to the optimal solution. The sensitivity of the optimal solution to the CSI estimation error is shown in Fig. 6.6. For very small σ_e, the optimal solution performs relatively well and the sum-rate increases with the number of users. However, as σ_e reaches and exceeds -1 dB, the performance degrades drastically as the number of users increases. This behavior is explained by the fact that, for a large number of users, it is unlikely that a user will have a large number of subcarriers allocated to it due to the increase of multiuser diversity. In this case, a reduced number of subcarriers with good CSI, and hence where a MCS with a high number of bits is used, will be allocated to each user. Consequently, an erroneous CSI with a relatively large error will lead to a significant decrease in the number of achievable bits for a given user. Conversely, for a small number of users, a relatively large number of subcarriers is allocated to the users. In this case, more robust MCS are used on the subcarriers (with lower number of bits), since the available power is subdivided among a large number of subcarriers. Consequently, an erroneous CSI on certain subcarriers will be partially compensated by the rate on the other subcarriers and hence will lead to less decrease in the number of achievable bits for a given user.

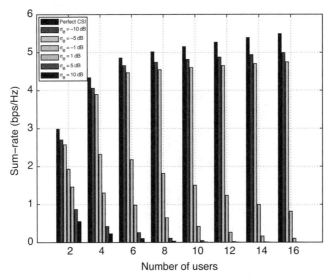

Figure 6.6 Impact of imperfect CSI on optimal resource allocation with discrete rates: case of channel coding and BER $= 10^{-6}$.

Comparing Algorithms 6.5 and 6.6 in both the perfect and imperfect CSI scenarios, it can be seen that their performance is comparable, and the additional complexity in Algorithm 6.6 is not justified, especially as the number of users increases. However, for a low number of users, Algorithm 6.6 has a relatively better performance. This behavior is explained by the fact that, for a large number of users, it is unlikely that a user will have a large number of subcarriers allocated to it due to the increase of multiuser diversity. In this case, the power is subdivided over a relatively small number of subcarriers and no considerable improvement can be made by Algorithm 6.6. Conversely, for a small number of users, a relatively large number of subcarriers is allocated to the users. In this case, when the user power is subdivided equally over the subcarriers, a relatively large amount of power is wasted compared to the optimal case. Hence, Algorithm 6.6 can ensure a more efficient use of the user power. Furthermore, allocating the remaining power in $P_{\text{Buffer},k}$ equally over the subcarriers provides a power margin that also makes Algorithm 6.6 less sensitive to imperfect CSI.

6.5.5 Comparison to Existing Algorithms

In this section, the proposed suboptimal algorithms in the discrete rates case are compared to the relevant literature, where most of the algorithms treat the problem of the transmit power minimization given per-user minimum rate constraints [18, 25, 30, 39]. This problem is the dual formulation of the problem investigated in this chapter, which consists of maximizing the sum-rate given a maximum power constraint. In Ref. [40], an algorithm proposed for power minimization is extended to the case of sum-rate maximization. The main focus in the literature is on downlink resource allocation [18, 25, 39, 40], with a few papers treating the uplink problem [30], where the power minimization problem is considered. These algorithms follow a three-steps approach to resource allocation with discrete rates: estimating the number of subcarriers to be allocated to each user, selecting and allocating the appropriate subcarriers, and using bit loading to allocate the power on these subcarriers.

In Ref. [25], subcarrier and transmit power allocation are considered in downlink OFDMA with fixed modulation per subcarrier. The QoS parameters, for example, target bit rate and bit error rate, are then used to determine the individual users' traffic demands. The powers needed to achieve a certain BER for each user over all subcarriers are grouped into a cost matrix, where the cost is the transmit power. The subcarrier allocation is then performed as follows:

- *Initial Allocation*: going through users sequentially, the unallocated subcarrier with lowest cost for the current user is allocated to that user. The process continues until all the subcarriers are allocated or the minimum rate requirements of all the users are satisfied.

- *Iterative Improvement*: To improve the initial allocation and reach better results with lower total transmit power from the BS, iterative subcarrier swapping is performed. Users are considered in pairs, and for each subcarrier allocated to a user, if power reduction (cost savings) can be obtained by swapping with a subcarrier of the second user, swapping is performed.

In Ref. [40], a downlink subcarrier, bit, and power allocation approach is presented for sum-rate maximization with minimum rate constraints. The algorithm of Ref. [40] is subdivided as follows:

- The number of subcarriers and an estimate of the transmit power to be allocated to each user are determined, based on the users' average channel gains and their rate requirements.

- Actual subcarrier allocation based on the channel gains of the users over all subcarriers is performed. The subcarrier assignment is based on a Hungarian algorithm.

- After subcarrier allocation, power is allocated by water-filling over the subcarriers assigned to each user.

In Ref. [30], an algorithm that performs OFDMA uplink resource allocation with the objective of minimizing the user transmit power while achieving a target user rate is presented. It can be summarized as follows:

- The minimum number of subcarriers to be allocated to each user is estimated such that the desired user rate can be reached with the user maximum uplink transmit power. The estimation of the number of subcarriers is based on the users' average channel-to-noise ratios (CNRs).

- The distribution of the subcarriers is performed as if the users choose the subcarriers alternatingly, with each user selecting the subcarrier having the best CNR. The user with the highest number of subcarriers begins, followed by the other users.

- Each user performs bit and power allocation via bit loading based on the sub-carrier allocation of the previous step.

A comprehensive survey of existing algorithms is presented in Ref. [9]. Selecting the best subalgorithm in the literature for each of the three steps, a new algorithm is developed in Ref. [9], where the focus is on the downlink power minimization problem. However, indications on extending the algorithm to the sum-rate maximization case are discussed based on the approach of Ref. [40], and indications on its application to the uplink scenario are presented following the approach of Ref. [30]. Following these guidelines, the algorithm below is presented and named as the Pietrzyk algorithm, after the author of Ref. [9]:

- *Step 1—Estimation of the Number of Subcarriers:* The number of subcarriers to be allocated to user k is estimated by

$$
N_k = \left\lfloor \left| \frac{\overline{g}_k}{\left(\sum_{m=1}^{K} \overline{g}_m \right)} \right| \right\rfloor
\tag{6.7}
$$

where $\lfloor \rfloor$ denotes the floor operation and $\overline{g}_k = \sum_{i=1}^{N} g_{k,i}/N$ is the average channel gain of user k. It should be noted that in the power minimization problem under a minimum rate constraint, the number of subcarriers is estimated as the minimum number of subcarriers required to achieve the target rate. To obtain the estimate, it is assumed that the highest allowed rate can be achieved on each subcarrier using the average channel gain $\overline{g}_k$ of each user.

- *Step 2—Allocation of Subcarriers to Users:* The users are sorted by decreasing order of the estimate obtained in Step 1. Then, an iterative process starts where a single subcarrier is allocated to each user per iteration. A user is allocated the available subcarrier to which it has the best CSI. The iterations are repeated until each user has received the number of subcarriers estimated in Step 1. If any subcarriers remain unassigned, each is allocated to the user having the best CSI on that subcarrier.

- *Step 3—Power Allocation on the Subcarriers:* Power is allocated by bit loading over the subcarriers. For each user k, the marginal power $\Delta P_{k,i}$ required to use the next available modulation level on subcarrier i is computed for all $i \in \mathcal{I}_{\text{sub},k}$. Then, $\Delta P_{k,i^*}$ is loaded on i^* such that $i^* = \arg \min_{i \in \mathcal{I}_{\text{sub},k}} \Delta P_{k,i}$. The process is repeated until no more power can be allocated without exceeding $P_{k,\max}$, the maximum transmit power. It should be noted that in the power minimization problem under a minimum rate constraint, the iterations are stopped when a user reaches its target minimum rate. Naturally, in the sum-rate maximization problem, the whole available power is used.

The results of comparing the proposed suboptimal algorithms (Algorithms 6.5 and 6.6) to the Pietrzyk algorithm are shown in Fig. 6.7 for the uncoded discrete rates case with a target BER of 10^{-6}. Algorithm 6.5 outperforms the Pietrzyk algorithm in

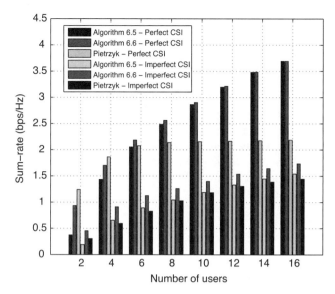

Figure 6.7   Comparison of the proposed algorithms to the Pietrzyk algorithm in the case of discrete rates with a target BER $= 10^{-6}$ and no channel coding.

the perfect CSI case when the number of users exceeds six, and in the imperfect CSI case with when the number of users is greater than two. Algorithm 6.6 outperforms the Pietrzyk algorithm in the perfect CSI case when the number of users exceeds four, and in all cases of the imperfect CSI scenario. Although power allocation by bit loading is the equivalent of water-filling in the continuous rates case, and hence should perform better than equal power allocation, the superiority of Algorithms 6.5 and 6.6 stems from their efficient subcarrier allocation rather than their power allocation. In fact, the uplink algorithms derived in the literature [30] and [9], are extensions of downlink algorithms. In the downlink, the estimation of the number of subcarriers to be allocated to each user before performing the actual allocation leads to efficient results, since the BS is the single source of power that will distribute the available power on the different users. Applying this step in the uplink, even after appropriate modifications, does not accurately take into account the distributed nature of the power constraint. Conversely, the proposed algorithms perform a good estimate of the optimal subcarrier allocation based on the continuous rates approximation. This approximation takes into account the gradual allocation of power on each subcarrier assigned to a given user, as reflected by (6.1) and Algorithm 6.4. When the number of users is small, both approaches lead to an approximately similar subcarrier allocation, and hence the superiority of bit loading over equal power allocation can be observed, as seen in Fig. 6.7 for the perfect CSI scenario. It should be noted that Algorithm 6.6 performs a good subcarrier allocation as Algorithm 6.5 does, and implements power allocation in a way similar to bit loading, although it starts from an initial approximation based on equal power. This explains its closer performance to the Pietrzyk algorithm for a small number of users in the perfect CSI scenario.

6.6   SUMMARY

In this chapter, low complexity suboptimal algorithms based on equal power allocation were proposed to approximate the discrete rates solution in addition to the continuous rates solution. It was shown that near optimal results can be achieved by the proposed algorithms for the continuous rates case. In the discrete rates scenario, results became closer to optimal as the number of modulation and coding schemes increased. Furthermore, equal power allocation was shown to be robust to imperfect channel state information conversely to the optimal solution, since equal power scheduling provides additional power on each subcarrier that can be used to compensate for estimation inaccuracies. In addition, the proposed algorithms compared favorably with the existing literature.

CHAPTER 7

SUBOPTIMAL IMPLEMENTATION WITH PROPORTIONAL FAIRNESS

Chapters 3 and 4 presented the optimal solutions for weighted ergodic sum-rate maximization with continuous and discrete rates, respectively. Chapter 5 extended the results of Chapters 3 and 4 to utility maximization in general. Chapter 6 presented suboptimal scheduling algorithms based on utility maximization. The algorithms were shown to approximate the optimal sum-rate maximization solutions of Chapters 3 and 4. In this chapter, suboptimal scheduling is implemented with utilities achieving fairness instead of considering only sum-rate maximization. Furthermore, the algorithms of Chapter 6 are extended to the practical scenario where subcarriers are grouped into blocks in order to reduce the scheduling complexity.

The chapter is organized as follows. Section 7.1 presents related background information and lists the contributions of this chapter. An overview of proportional fair (PF) scheduling and its equivalence to the Nash bargaining solution (NBS) are presented in Section 7.2. The proposed scheduling algorithms are described in Section 7.3. A discussion of the utilities ensuring proportional fair scheduling is presented in Section 7.4. The simulation results are presented and analyzed in Section 7.5. The chapter summary is presented in Section 7.6.

7.1 BACKGROUND

When the number of subcarriers increases, the scheduling complexity increases, even with linear complexity algorithms, since the complexity is dependent on the number of subcarriers. Hence, resource allocation is not performed on a subcarrier by subcarrier basis in state-of-the-art OFDMA-based systems. Instead, subcarriers are assembled into groups of consecutive subcarriers, called resource blocks (RBs) in LTE for example, and allocation is performed on an RB basis. In LTE, the available spectrum is divided into RBs consisting of 12 adjacent subcarriers, allocated in a 0.5 ms time slot. The shortest assignment unit consists of two consecutive slots, that is, for a duration of 1 ms, which is the duration of one transmission time interval (TTI) [41, 83, 116]. In WiMAX, 192 data OFDM subcarriers are distributed in 16 subchannels of 12 subcarriers each. Each subchannel is made of four groups of three adjacent subcarriers each [42].

Resource Allocation in Uplink OFDMA Wireless Systems: Optimal Solutions and Practical Implementations,
Elias E. Yaacoub and Zaher Dawy.
© 2012 by the Institute of Electrical and Electronics Engineers, Inc. Published 2012 by John Wiley & Sons, Inc.

In Chapter 6, suboptimal scheduling algorithms for sum-rate maximization in the continuous and discrete rates scenarios were presented. Algorithm 6.4 was shown to perform an approximation to scheduling with discrete rates by using continuous rates and an SNR gap. Hence, in this chapter, two suboptimal algorithms that extend Algorithm 6.4 are presented, in order to accommodate RB-based scheduling in addition to utility maximization. In this case, subcarrier-based scheduling can be seen as a special case of RB-based scheduling where each RB consists of a single subcarrier. In addition, sum-rate maximization can be seen as a special case of utility maximization where the utility consists of the rate itself, not another function of the rate.

The proposed algorithms, Algorithm 7.1 and Algorithm 7.2, both have polynomial complexity. However, Algorithm 7.1 has a higher complexity than Algorithm 7.2. Algorithm 7.2 constitutes a direct extension of Algorithm 6.4 to general utility scheduling using RBs instead of individual subcarriers. However, Algorithm 7.1 is derived in order to provide a fair comparison with other algorithms in the literature that have the same complexity order (quadratic complexity). Algorithm 7.1 will be used in the results of Chapter 8, since it is shown in this chapter to achieve the best performance and its extension to scheduling in a distributed base station scenario has a low complexity. However, Algorithm 7.2 is shown in this chapter to have a comparable performance to Algorithm 7.1 with a reduced complexity. Hence, Algorithm 7.2, or direct extensions of it, will be used in the scheduling results presented in Chapters 9–13.

After presenting the utility maximization algorithms, the main emphasis in this chapter will be on utilities achieving proportional fairness. The importance of PF scheduling in providing fairness is explained by resorting to a game theoretical interpretation. The Nash bargaining problem (NBP) [44] is a well-known scenario in game theory. Players in the NBP negotiate to maximize their payoffs, given a set of shared resources. The optimal solution of the NBP, the NBS, consists of distributing the resources in a way to maximize the product of the payoffs [45]. It was shown that PF scheduling is equivalent to the implementation of the NBS in the resource allocation of wireless communication systems, the payoff of each user being its rate [43, 46]. PF scheduling is widely investigated in the literature, mainly in the framework of centralized resource allocation [19, 29, 117]. With OFDMA adopted as the accessing scheme of next generation cellular systems, for example, 3GPP LTE and mobile WiMAX (IEEE 802.16e), several applications of PF to OFDMA were studied [47–49].

Hence, the main topics discussed in this chapter can be summarized as follows:

1. Presenting low-complexity suboptimal scheduling algorithms extending the results of Chapter 6 in order to accommodate utility maximization with RB-based scheduling.

2. Applying the proposed algorithms with various utility functions with an emphasis on utilities achieving proportional fairness.

3. Comparing the proposed algorithms to other algorithms in the literature and showing that Algorithm 7.1 achieves a better performance at the same complexity whereas Algorithm 7.2 achieves a comparable performance with less complexity.

7.2 PROPORTIONAL FAIR SCHEDULING

This section describes PF scheduling and its relation to the NBS.

7.2.1 PF Scheduling Methods

In Ref. [118], proportional fairness was defined as follows: a feasible rate vector $\mathbf{R}$ is proportional fair if for any other feasible rate vector $\mathbf{R}^*$, the aggregate of proportional changes satisfies

$$\sum_{k=1}^{K} \frac{R_k^* - R_k}{R_k} \leq 0 \tag{7.1}$$

where the summation is over the number of users K.

In CDMA systems, proportional fairness consists of allocating all the available resources to a single user for a given TTI. The user k^* selected is the one satisfying [119]

$$k^* = \arg\max_k R_k^{(n)}/D_{k,\text{tot}} \tag{7.2}$$

with $R_k^{(n)}$ the rate achievable at TTI n and $D_{k,\text{tot}}$ the total achieved rate in a previous time window of fixed duration, for example, the last 1000 TTIs. This approach is known as proportional fairness in time (PFT).

Applying the previous scheme in OFDMA consists of allocating all the subcarriers to a single user at a given scheduling interval. This solution is inefficient, since subcarrier allocation plays an important role in the scheduling process in OFDMA. Hence, at a given TTI n, (7.2) is applied on a subcarrier basis, that is, subcarrier i is allocated to user k^* satisfying [48]

$$k^* = \arg\max_k R_{k,i}^{(n)}/D_{k,\text{tot}} \tag{7.3}$$

with $R_{k,i}^{(n)}$ the rate achievable by user k over subcarrier i at TTI n. A more accurate approach would be to allocate subcarrier i to user k^* satisfying [48]

$$k^* = \arg\max_k \frac{R_{k,i}^{(n)}}{D_{k,\text{tot}} + \sum_{j \in \mathcal{I}_{\text{sub},k}^{(n)}, j \neq i} R_{k,j}^{(n)} T} \tag{7.4}$$

with $\mathcal{I}_{\text{sub},k}^{(n)}$ the set of subcarriers already allocated to user k during the scheduling process for TTI n, and T is the time duration of a TTI. The difference between (7.3) and (7.4) is that in (7.4), the rates on the subcarriers already allocated to a user at the current scheduling interval are taken into account before allocating the remaining subcarriers at the same interval.

Scheduling according to (7.3) and (7.4) is known as proportional fairness in time and frequency (PFTF) since it involves frequency scheduling on a per subcarrier basis while taking into account the achieved data rate at previous TTIs.

Considering, in (7.4), the current TTI only and neglecting the rate achieved from previous allocations, then subcarrier i is allocated to user k^* satisfying [120]

$$k^* = \arg\max_k \frac{R_{k,i}^{(n)}}{\displaystyle\sum_{j \in \mathcal{I}_{\text{sub},k}^{(n)}, j \neq i} R_{k,j}^{(n)} T} \tag{7.5}$$

The approach of (7.5) is referred to as proportional fairness in frequency (PFF) since the time dependence is neglected and only the frequency dimension is used in the resource allocation process. In the sequel, the superscript (n) will be dropped to simplify the notations when no confusion can occur.

It was shown in Ref. [118] that maximizing the logarithmic utility, that is,

$$\max \sum_{k=1}^{K} \ln(R_k) \tag{7.6}$$

achieves proportional fairness. This result is widely used in the literature [19, 47, 117]. Using the achievable rate at each scheduling instant in (7.6) achieves PFF [117], whereas including the previous scheduling instants by using the average rate achieves PFTF [49].

Theorem 7.1. *PFF allocates at least one subcarrier to each user as long as the number of users does not exceed the number of subcarriers.*

Proof. Let K be the number of users and N the number of subcarriers. From (7.6), it is clear that to avoid a $-\infty$ value, no user should have a zero rate. Hence, PFF allocates at least one subcarrier to each user when $K \leq N$. When $K = N$, exactly one subcarrier is allocated to each user. However, when $K > N$, the K subcarriers are allocated to the K users having the best channel conditions (one for each), and there are no sufficient resources for the remaining users. □

It should be noted that the case $K > N$ does not represent a problem for PFTF where the time dimension is used in the scheduling process, in addition to the frequency and multiuser diversity dimensions. Hence, the users alternate on the available subcarriers so that the average allocation is fair.

7.2.2 Equivalence of PF and NBS

In this section, in order to confirm the equivalence of PF and NBS, the resource allocation problem in the OFDMA uplink is modeled as a bargaining game. Each user is considered to be a player (in this section, both terms user and player are used interchangeably) who wants to maximize its payoff, considered to be its rate.

Cooperation is assumed between players. Consequently, players should share the resources in an optimal way, that is, a way they cannot jointly improve on. The resources to be shared are the subcarriers in each TTI. Allocating the shared resources in a way to maximize the users' payoffs is equivalent to allocating subcarriers to users in a way to maximize each user's rate, given the shares allocated to the other users. It is a well-known result in game theory that the solution to the cooperative bargaining problem maximizes the Nash product N_P [45]:

$$N_P = \prod_{k=1}^{K} \left(W_k(x_k) - F_k \right) \tag{7.7}$$

where x_k represents the fraction of resources allocated to player k, $W_k(x_k)$ corresponds to the payoff of player k when x_k is allocated to it, and F_k is the payoff of player k in the case where no agreement is reached in the bargaining problem. The NBS is illustrated in Fig. 7.1.

In the OFDMA scheduling problem, the player payoff is the rate achieved, that is, $W_k(x_k) = R(P_k, \mathcal{I}_{\mathrm{sub},k})$, with P_k the total transmission power of user k and $\mathcal{I}_{\mathrm{sub},k}$ the set of subcarriers allocated to user k. In addition, $F_k = 0$ since no transmission occurs if no agreement on subcarrier allocation is reached. Hence, the optimization problem becomes

$$\max \prod_{k=1}^{K} R(P_k, \mathcal{I}_{\mathrm{sub},k}) \tag{7.8}$$

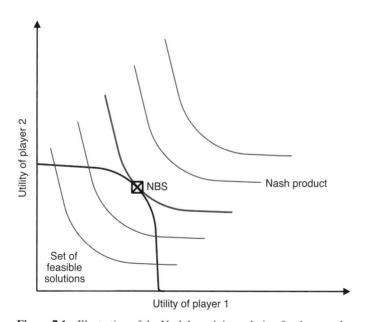

Figure 7.1   Illustration of the Nash bargaining solution for the two players case.

TABLE 7.1 Analogy Between Game Theory and Resource Allocation in OFDMA Systems

Game theory	Resource allocation
Player	User
Resources	Subcarriers
Payoff W_k	Rate R_k
Payoff with no agreement F_k	Rate with no agreement $R_k = 0$
Nash bargaining game	Resource allocation problem
Nash product	Rate product
Nash bargaining solution	Proportional fair scheduling

Since the logarithm is a continuous strictly increasing function, solving the problem in (7.8) is equivalent to finding the solution of the following problem:

$$\ln\left(\max\left(\prod_{k=1}^{K} R(P_k, \mathcal{I}_{\text{sub},k}) \right) \right)$$
$$= \max \ln\left(\prod_{k=1}^{K} R(P_k, \mathcal{I}_{\text{sub},k}) \right) \tag{7.9}$$
$$= \max \sum_{k=1}^{K} \ln\left(R(P_k, \mathcal{I}_{\text{sub},k}) \right)$$

Hence, this proves that the logarithm of the rate, when used as a utility, ensures proportional fairness, since maximizing it leads to maximizing the Nash product thus leading to the Nash bargaining solution. The analogy between game theory and proportional fair resource allocation in the uplink of OFDMA systems is described in Table 7.1.

7.3 LOW COMPLEXITY UTILITY MAXIMIZATION ALGORITHMS

In this section, two low complexity scheduling algorithms are presented. These algorithms perform subcarrier allocation in a centralized scenario within the OFDMA uplink framework. The algorithms are presented in decreasing order of complexity. The performance of the two algorithms is compared in Section 7.5 via simulations. The proposed algorithms are channel aware scheduling algorithms that assume the knowledge of the CSI at the BS. It should be noted that the algorithms can be used with any utility function and have the property of ensuring proportional fairness when the utility used is the logarithm of the rate.

Throughout this book, N is used as the number of subcarriers. However, to avoid any confusion when scheduling is performed on an RB basis, N_{RB} is used to represent the number of RBs and N_{sub} to represent the number of subcarriers.

$\mathcal{I}_{\text{RB},k}$ is used to denote the set of RBs allocated to user k, $\mathcal{I}_{\text{sub},k}$ the set of subcarriers allocated to user k, P_k the instantaneous transmission power of user k,

$P_{k,\max}$ its maximum transmission power, and R_k its achievable rate. $U(R_k|\mathcal{I}_{\mathrm{RB},k})$ is the utility of user k as a function of the rate R_k given the allocation $\mathcal{I}_{\mathrm{RB},k}$. The utility function depends on the rate, and could vary depending on the different services and QoS requirements. Letting the utility equal to the rate, the proposed algorithms lead to a maximization of the sum-rate of the cell, whereas the logarithmic utility function is associated with the proportional fairness for the utility-based optimization.

The proposed algorithms consist of allocating RB i to user k in a way to maximize the difference

$$\Lambda_{k,i} = U_k(R_k|\mathcal{I}_{\mathrm{RB},k} \cup \{i\}) - U_k(R_k|\mathcal{I}_{\mathrm{RB},k}) \tag{7.10}$$

where the marginal utility, $\Lambda_{k,i}$, represents the gain in the utility function U_k when RB i is allocated to user k, compared to the utility of user k before the allocation of i.

Algorithm 7.1

> **Quadratic Complexity Scheduling Algorithm**
> **Subcarrier Allocation:**
> **for** k **such that** $1 \leq k \leq K$
> $\mathcal{I}_{\mathrm{sub},k} \leftarrow \emptyset$
> **end for**
> **while** $\mathcal{I}_{\mathrm{avail\_RB}} \neq \emptyset$
>   **for all** i **such that** $i \in \mathcal{I}_{\mathrm{avail\_RB}}$
>   **for all** k **such that** $1 \leq k \leq K$
>   **Compute** $U_k(P_k, \mathcal{I}_{\mathrm{RB},k} \cup \{i\})$
>   **if** $\mathcal{I}_{\mathrm{RB},k} \neq \emptyset$
>   **Compute** $U_k(P_k, \mathcal{I}_{\mathrm{RB},k})$
>   **else**
> $U_k(P_k, \mathcal{I}_{\mathrm{RB},k}) \leftarrow 0$
>   **end if**
> $\Lambda_{k,i} \leftarrow U_k(P_k, \mathcal{I}_{\mathrm{RB},k} \cup \{i\}) - U_k(P_k, \mathcal{I}_{\mathrm{RB},k})$
>   **end for**
>   **end for**
> $(k^*, i^*) \leftarrow \arg \max_{(k,i)} \Lambda_{k,i}$
>   **if** $\Lambda_{k^*,i^*} > 0$
> $\mathcal{I}_{\mathrm{RB},k^*} \leftarrow \mathcal{I}_{\mathrm{RB},k^*} \cup \{i^*\}$
>   **end if**
> $\mathcal{I}_{\mathrm{avail\_RB}} \leftarrow \mathcal{I}_{\mathrm{avail\_RB}} \setminus \{i^*\}$
> **end while**
> **Power Allocation:**
> **Apply Algorithm   6.2 for power allocation**

Algorithm 7.2

> **Linear Complexity Scheduling Algorithm**
> **Subcarrier Allocation:**
> **for** k **such that** $1 \leq k \leq K$

$\mathcal{I}_{\text{RB},k} \leftarrow \emptyset$
end for
for i **such that** $1 \leq i \leq N_{\text{RB}}$
   for all k **such that** $1 \leq k \leq K$
      Compute $U_k(P_k, \mathcal{I}_{\text{RB},k} \cup \{i\})$
      if $\mathcal{I}_{\text{RB},k} \neq \emptyset$
         Compute $U_k(P_k, \mathcal{I}_{\text{RB},k})$
      else
         $U_k(P_k, \mathcal{I}_{\text{RB},k}) \leftarrow 0$
      end if
      $\Lambda_{k,i} \leftarrow U_k(P_k, \mathcal{I}_{\text{RB},k} \cup \{i\}) - U_k(P_k, \mathcal{I}_{\text{RB},k})$
   end for
   $k^* \leftarrow \arg\max_k \Lambda_{k,i}$
   if $\Lambda_{k^*,i} > 0$
      $\mathcal{I}_{\text{RB},k^*} \leftarrow \mathcal{I}_{\text{RB},k^*} \cup \{i\}$
   end if
end for
Power Allocation:
Apply Algorithm   6.2 for power allocation

The following lemmas highlight some important properties of the proposed algorithms.

Lemma 7.1. *The proposed algorithms achieve proportional fairness in frequency when the utility considered is the logarithm of the rate at the current scheduling TTI.*

Proof. Let $C_j^{(n)}$ be the rate of a user at TTI n when a given RB j is allocated to it and $C^{(n)}$ its rate at TTI n without RB j. When the utility is the user rate, the scheduler allocates j to the user that maximizes the difference $(C_j^{(n)} - C^{(n)})$, which is the user having the best channel conditions, that is, in most cases the user closest to the BS. On the other hand, when the utility is the logarithm of the rate, the scheduler allocates j to the user that maximizes the difference $(\ln(C_j^{(n)}) - \ln(C^{(n)}))$, or, equivalently, the ratio $C_j^{(n)}/C^{(n)}$, thus achieving (7.5) by allocating j to the user that achieves the largest proportional increase in the rate. $\square$

Lemma 7.2. *The proposed algorithms achieve proportional fairness in time and frequency when the utility considered is the logarithm of the average rate of each user, that is, when the rate achieved at previous TTIs is included in the scheduling process.*

Proof. The proof is similar to that of Lemma 7.1. In fact, letting $C = D_{\text{tot}} + C^{(n)}$ and $C_j = D_{\text{tot}} + C_j^{(n)}$, the scheduler maximizes the difference $(\ln(C_j) - \ln(C))$, or, equivalently, the ratio C_j/C, which corresponds to achieving (7.3). $\square$

The proof of Lemmas 7.1 and 7.2 used the fact that the algorithms maximize the marginal utility. However, the use of the difference in (7.10) is not only applicable to PF but also leads to better results regardless of the utility (e.g., sum-rate utility) as shown in the following theorem.

Theorem 7.2. *Marginal utility is capable of achieving better performance than full utility, that is, finding the user k_1^* that maximizes the marginal utility $\Lambda_{k,j}$ leads to a sum-utility greater than finding the user k_2^* that maximizes the full utility $U(R_k | \mathcal{I}_{\text{sub},k} \cup \{j\})$.*

Proof. As a proof outline, the following case is considered: $U(R_{k_1^*} | \mathcal{I}_{\text{sub},k_1^*} \cup \{j\}) = X - \varepsilon$, $U(R_{k_1^*} | \mathcal{I}_{\text{sub},k_1^*}) = X/2$, $U(R_{k_2^*} | \mathcal{I}_{\text{sub},k_2^*} \cup \{j\}) = X + \varepsilon$, and $U(R_{k_2^*} | \mathcal{I}_{\text{sub},k_2^*}) = X$, with $X \gg \varepsilon$. In this case, the proposed algorithm allocates j to k_1^*, leading to a sum-utility increase of $\Lambda_{k_1^*,j} = X/2 - \varepsilon$. On the other hand, an algorithm maximizing the full utility would allocate j to k_2^* since $U(R_{k_2^*} | \mathcal{I}_{\text{sub},k_2^*} \cup \{j\}) = X + \varepsilon$ is greater than $U(R_{k_1^*} | \mathcal{I}_{\text{sub},k_1^*} \cup \{j\}) = X - \varepsilon$. However, in this case, the increase in the sum-utility by allocating subcarrier j to user k_2^* is only $U(R_{k_2^*} | \mathcal{I}_{\text{sub},k_2^*} \cup \{j\}) - U(R_{k_2^*} | \mathcal{I}_{\text{sub},k_2^*}) = \varepsilon$. Since $X/2 - \varepsilon > \varepsilon$, the maximization of marginal utility leads to better results. $\quad\square$

7.3.1 Complexity Analysis of the Utility Maximization Algorithms

Algorithm 7.1 allocates each RB after performing a linear search on the users and RBs in order to find the user–RB pair that maximizes the marginal utility. Hence, the complexity to allocate the first RB is $\mathcal{O}(N_{\text{RB}} K)$, the complexity to allocate the second RB is $\mathcal{O}((N_{\text{RB}} - 1)K)$, and so on. Consequently, the total complexity of the algorithm is

$$\mathcal{O}\left(N_{\text{RB}} K + (N_{\text{RB}} - 1)K + \cdots + 2K + K\right) = \mathcal{O}\left(\frac{N_{\text{RB}}(N_{\text{RB}} + 1)}{2} K\right)$$
$$\approx \mathcal{O}(N_{\text{RB}}^2 K) \quad (7.11)$$

that is, the algorithm has linear complexity in the number of users and quadratic complexity in the number of RBs, and thus could be easily implemented in real-time. Algorithm 7.2 differs from Algorithm 7.1 in that the scheduler is not looking for an RB–user pair. Instead, it goes through the RBs sequentially, that is, considers only the first available RB, and allocates the selected RB to the user that achieves to the maximum marginal utility when the RB is allocated to it. Hence, Algorithm 7.2 allocates each RB after performing a linear search on the users in order to find the user that maximizes the marginal utility. Consequently, the total complexity of the algorithm is $\mathcal{O}(N_{\text{RB}} K)$, that is, the algorithm has linear complexity in the number of users and in the number of RBs, and thus could be implemented in real-time, faster than Algorithm 7.1.

7.3.2 Comparison to Existing Algorithms

Optimal subcarrier allocation consists of performing a combinatorial search of complexity $\mathcal{O}(K^{N_{RB}})$, which is prohibitive when the number of users and/or RBs increases. Therefore, suboptimal algorithms are developed in the literature. In Refs [95, 121], an uplink scheduling algorithm for OFDMA is presented. It is used to compare the sum-rate performance between water-filling and equal power allocation, and reach the conclusion that both schemes lead to comparable results. It allocates subcarrier i^* to user k^* such that $(k^*, i^*) = \arg\max_{k,i\in\mathcal{I}_{avail\_sub}} g_{k,i}$. It does not consider the grouping of subcarriers into RBs, and does not incorporate the effect on the rate achieved over the other subcarriers in $\mathcal{I}_{sub,k^*}$ after adding i^* to the set. Consequently, assuming this algorithm is extended to the case where subcarriers are grouped into RBs, it still does not capture all the details as in Algorithm 7.1, which will consequently lead to better performance while having a similar complexity $\mathcal{O}(N_{RB}^2 K)$. In Refs [111, 117], an uplink scheduling algorithm that extends the algorithm of Ref. [95] to RB uplink scheduling with equal power allocation per RB is presented. The algorithm of Ref. [117] can be described as follows:

- Perform an initial search over all user–RB pairs ($\mathcal{O}(N_{RB}^2 K)$ complexity) to find $(k_1^*, i^*) = \arg\max_{k,i\in\mathcal{I}_{avail\_RB}} g_{k,i}$.
- After this initial step, an RB i^* to be allocated is determined.
- Then, a search similar to the one performed in Algorithm 7.2 is used to determine the user k^* to which i^* is allocated ($\mathcal{O}(N_{RB} K)$ complexity).

Obviously, performing the initial search over the difference in (7.10) as in Algorithm 7.1 will lead to better results since no RBs are eliminated from the search. Consequently, Algorithm 7.1 will lead to better results with less complexity. Nevertheless, since the algorithm of [111] constitutes an extension of the algorithm of Ref. [95] to the RB case, its performance will be compared to the two algorithms presented in this chapter. This algorithm will be denoted in the simulations by Myung's algorithm.

In Ref. [47], an algorithm that can be summarized as follows is presented:

- The OFDMA uplink scheduling algorithm finds a user–subcarrier pair (k^*, i^*) that has the highest marginal utility defined in (7.10) (with $\mathcal{I}_{RB}$ substituted by $\mathcal{I}_{sub}$) among all available subcarriers and users.
- Then, the algorithm allocates i^* to k^*.
- However, when determining the utility as in (7.10), the algorithm of Ref. [47] determines $P_{k,i}$ by subdividing the power P_k among all unallocated subcarriers in addition to the ones allocated to user k, using water-filling.

To compare this algorithm to the algorithms presented in this chapter, it is extended to RB scheduling with equal power allocation as follows: scheduling is performed on an RB-by-RB basis instead of a subcarrier-by-subcarrier basis, and equal power allocation is applied (instead of water-filling) over all unallocated subcarriers in addition the ones allocated to a given user. This algorithm will be denoted in the simulations by Ng's algorithm. Clearly, its complexity is $\mathcal{O}(N_{RB}^2 K)$, similarly to Algorithm 7.1.

In Ref. [96], an OFDMA uplink scheduling algorithm that maximizes the sum-rate is presented. It can be summarized as follows:

- The algorithm allocates subcarriers using a search similar to the one performed in Algorithm 7.2, with the utility being limited to the user rate and with $\mathcal{I}_{RB}$ being substituted by $\mathcal{I}_{sub}$.

- After each allocation, the algorithm checks if the user has achieved its target data rate.

- If that user has received enough subcarriers to achieve its target rate, it is excluded from the remainder of the subcarrier allocation process, that is, none of the remaining subcarriers are allocated to it, unless all other users have reached their target rates.

Thus, the algorithm of Ref. [96] tries to ensure some fairness by preventing subcarriers from being allocated to a given user after it reaches its target data rate, unless some unallocated subcarriers remain after all users achieve their target rates. Since Algorithm 7.2 constitutes a generalization of this algorithm to any utility maximization, ensures fairness via appropriate utility selection without the introduction of target rates, and applies equal power allocation over RBs instead of water-filling over the subcarriers, the algorithm of Ref. [96] will not be considered in the simulations.

7.3.3 Rate Calculations

Although RB-based scheduling is adopted, the user rate in OFDMA consists of the sum of the rate on each RB, which is equivalent to the sum of the rate on each individual subcarrier, and subdividing the power equally among RBs is equivalent to equal power allocation over subcarriers. Hence, for the rate calculations, the following expression is considered:

$$R(P_k, \mathcal{I}_{sub,k}) = \sum_{i=1}^{N_{sub}} \alpha_{k,i} \frac{B}{N_{sub}} \cdot \log_2(1 + \beta \gamma_{k,i}) \tag{7.12}$$

where B is the total bandwidth, $\alpha_{k,i} = 1$ if subcarrier i is part of an RB allocated to user k (i.e., $i \in \mathcal{I}_{sub,k}$), and β is the SNR gap given by (6.3).

Each user is assumed to transmit at the maximum power ($P_k = P_{k,max}$), and the power is assumed to be subdivided equally among all the subcarriers allocated to that user, as discussed in Chapter 6. Hence, the SNR over a single subcarrier, $\gamma_{k,i}$, is given by

$$\gamma_{k,i} = \frac{P_{k,i} H_{k,i}}{\sigma_i^2} \tag{7.13}$$

where $H_{k,i}$ is the channel gain over subcarrier i allocated to user k, σ_i^2 is the noise power, and $P_{k,i}$ is the power allocated by user k over subcarrier i. It is given by

$$P_{k,i} = \frac{P_k}{|\mathcal{I}_{sub,k}|} \tag{7.14}$$

7.4 PROPORTIONAL FAIR UTILITIES

The NBS in cooperative game theory assumes the presence of some authority to enforce the agreement between the different players [45]. In a centralized scheduling scenario, no negotiations take place between the players. The BS is responsible for making the scheduling decisions. However, since the NBS in a bargaining scenario is equivalent to proportional fairness, it could be assumed that *virtual* negotiations take place between the players that reach the optimal NBS, and that the BS is the authority assumed to enforce the cooperative solution.

To investigate the performance of the proposed algorithm in the centralized case, the following schemes are considered:

- *PFF*: This corresponds to using, in the proposed algorithms, $U_k = \ln(R_k^{(n)})$. It can be easily shown that using this utility in (7.10) is equivalent to applying (7.5).

- *PFTF1*: This corresponds to using, in the algorithm, $U_k = \ln\left(\sum_n R_k^{(n)}\right)$. It can easily be shown that using this utility in (7.10) is equivalent to applying (7.6) with average rate.

- *PFTF2*: This corresponds to using, in the proposed algorithms,

$$
U_k = \frac{\sum\limits_{j \in \mathcal{I}_{\text{sub},k}^{(n)}} R_{k,j}^{(n)} T}{D_{k,\text{tot}} + \sum\limits_{j \in \mathcal{I}_{\text{sub},k}^{(n)}, j \neq i} R_{k,j}^{(n)} T}
$$

It can easily be shown that using this utility in (7.10) is equivalent to applying (7.4).

- *PFTF3*: This corresponds to using, in the proposed algorithms,

$$
U_k = \sum\limits_{j \in \mathcal{I}_{\text{sub},k}^{(n)}} R_{k,j}^{(n)} T / D_{k,\text{tot}}
$$

It can easily be shown that using this utility in (7.10) is equivalent to applying (7.3).

- *Greedy Scheduling*: This corresponds to using, in the algorithm, $U_k = R_k^{(n)}$. This scheme maximizes the sum-rate and is unfair to edge users. It is considered for comparison purposes.

PFT is not used since it is known to perform worse than the other PF schemes [48]. PFTF1, PFTF2, and PFTF3 are variants of the same objective function, with different degrees of accuracy. They are presented to test the performance of the algorithm in the different scenarios and to select the scheme achieving the best results. The selected PFTF scheme represents a benchmark to which the distributed scheduling results of Chapter 9 will be compared.

7.5   RESULTS AND DISCUSSION

7.5.1   Simulation Model

The simulation model consists of a single cell of radius $R_c = 1$ km with a BS equipped with an omnidirectional antenna. The total bandwidth considered is $B = 5$ MHz. A target BER of 10^{-6}, and a TTI duration of 1 ms are considered. The maximum mobile transmit power is considered to be 125 mW. All mobiles are assumed to transmit at the maximum power, and the power is subdivided equally among all subcarriers allocated to the mobile.

The channel gain over subcarrier i corresponding to user k is given by

$$H_{k,i,\mathrm{dB}} = (-\kappa - \upsilon \log_{10} d_k) - \xi_{k,i} + 10 \log_{10} F_{k,i} \qquad (7.15)$$

In (7.15), the first factor captures propagation loss, with κ a constant chosen to be 128.1 dB, d_k the distance in km from mobile k to the BS, and υ the path loss exponent, which is set to a value of 3.76. The second factor, $\xi_{k,i}$, captures log-normal shadowing with an 8 dB standard deviation, whereas the last factor, $F_{k,i}$, corresponds to Rayleigh fading with a Rayleigh parameter b such that $E[b^2] = 1$. Unless otherwise specified, Rayleigh fading is assumed to be IID over each of the subcarriers. Perfect CSI estimation is assumed at the BS.

7.5.2   PFF and PFTF Utility Comparison

In this section, eight RBs are considered, with one subcarrier per RB. The users are assumed to be located at fixed distances equally spaced from the cell center to the cell edge. Scheduling is performed using Algorithm 7.2. It should be noted that the number of users and subcarriers is relatively low in order to visualize the performance for the various user locations in Figs 7.2–7.4. The algorithm will be applied in other scenarios in the next section.

To illustrate the performance of the different PF schemes in the centralized scheduling case, three scheduling scenarios are considered:

- *Centralized Scenario 1*: The number of users is less than the number of subcarriers. This scenario is shown in Fig. 7.2 for four users and eight subcarriers. The sum-utility results are displayed in Table 7.2.

- *Centralized Scenario 2*: The number of users is equal to the number of subcarriers. This scenario is shown in Fig. 7.3 for eight users and eight subcarriers. The sum-utility results are displayed in Table 7.3.

- *Centralized Scenario 3*: The number of users is greater than the number of subcarriers. This scenario is shown in Fig. 7.4 for 16 users and eight subcarriers. The sum-utility results are displayed in Table 7.4.

Figs 7.2 and 7.3 show that the three PFTF schemes outperform the PFF scheme in terms of the achieved rate for all user positions. However, Fig. 7.4 shows that the rate of PFF is higher for the nearest eight users. Figs 7.2–7.4 show that all three PFTF schemes tend to allocate on average an approximately equal number of subcarriers

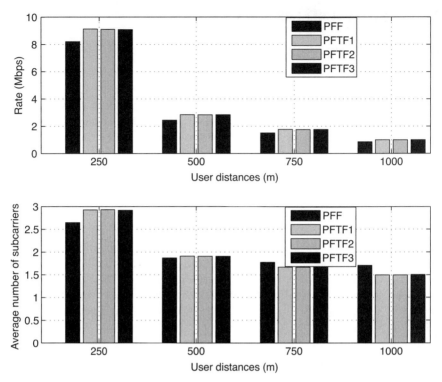

Figure 7.2 User rate and subcarrier allocation results as a function of the distance from the base station: four users and eight subcarriers. This figure is reprinted, with permission from Elsevier, from E. Yaacoub and Z. Dawy, "Achieving the Nash Bargaining Solution in OFDMA Uplink Using Distributed Scheduling with Limited Feedback", *International Journal of Electronics and Communication AEU (Elsevier)*, 65(4), 320–330, 2011 ©2011 Elsevier.

to each user, and this trend becomes clearer as the number of users increases. In this regard, the PFF scheme outperforms the PFTF schemes when the number of subcarriers is greater than or equal to the number of users. In fact, when the number of subcarriers is equal to the number of users, Fig. 7.3 shows that the PFF scheme allocates exactly one subcarrier to each user, as discussed in Theorem 7.1. When the number of users increases, it can be seen from Fig. 7.4 that the PFF scheme allocates most resources to the nearest users, since instantaneous subcarrier allocation mandates providing N_{sub} subcarriers to the N_{sub} (out of $K > N_{\text{sub}}$) users having the best channel conditions, because the PFF scheme does not keep track of the previously achieved data rate of a given user.

From Figs 7.2–7.4, it can be seen that the PFTF schemes have a close performance, and it is difficult to determine which scheme performs best. However, from Tables 7.2–7.4, it can be concluded that PFTF1 achieves the best performance in all three scenarios. PFTF2 outperforms PFTF3 in the 8 and 16 users scenarios, but they have an almost equal performance with 4 users (with a negligible superiority for PFTF3). All three PFTF schemes outperform PFF in terms of the PF utilities (product

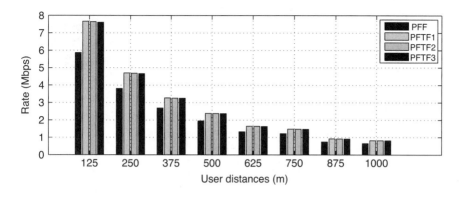

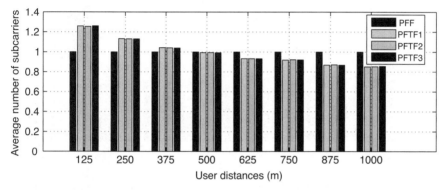

Figure 7.3 User rate and subcarrier allocation results as a function of the distance from the base station: eight users and eight subcarriers. This figure is reprinted, with permission from Elsevier, from E. Yaacoub and Z. Dawy, "Achieving the Nash Bargaining Solution in OFDMA Uplink Using Distributed Scheduling with Limited Feedback", *International Journal of Electronics and Communication AEU (Elsevier)*, 65(4), 320–330, 2011 ©2011 Elsevier.

and sum of logarithms) in all cases, but also in the sum-rate utility when the number of users is less than or equal to the number of subcarriers (Tables 7.2 and 7.3). As the number of users increases, the performance gap between PFF and the PFTF schemes increases, and the behavior of PFF becomes closer to greedy scheduling, as seen from Table 7.4. It should be noted that the rate used in the product utility is in Mbps in order to avoid excessively large numbers, whereas the rate used in the logarithmic utility is in bps.

7.5.3 RB-based Scheduling: Greedy and PFF Utilities

In this section, Algorithm 7.2 is implemented for RB-based scheduling. The $B = 5$ MHz bandwidth is subdivided into 25 RBs of 12 subcarriers each in this section in addition to Sections 7.5.4 and 7.5.5. The users are distributed in the cell according to a uniform distribution. Results from the round robin (RR) algorithm are presented for reference. The implementation of RR is straightforward. The number of RBs is fixed, and the RBs are allocated to users on a first-come first-served basis.

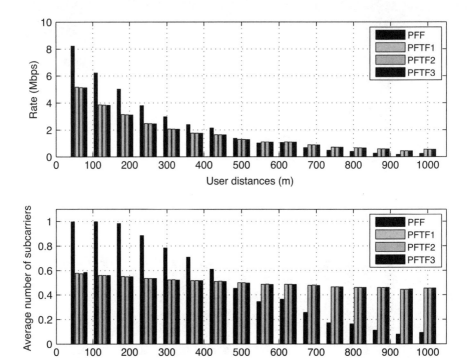

Figure 7.4 User rate and subcarrier allocation results as a function of the distance from the base station: 16 users and eight subcarriers. This figure is reprinted, with permission from Elsevier, from E. Yaacoub and Z. Dawy, "Achieving the Nash Bargaining Solution in OFDMA Uplink Using Distributed Scheduling with Limited Feedback", *International Journal of Electronics and Communication AEU (Elsevier)*, 65(4), 320–330, 2011 ©2011 Elsevier.

Fig. 7.5 shows the cell sum-rate results. The results corresponding to round robin will be denoted by "RR", and those corresponding to Algorithm 7.2 with $U = R$ and $U = \ln(R)$ (instantaneous rate) will be denoted by "Greedy" and "PFF", respectively. It can be seen from Fig. 7.5 that Algorithm 7.2 outperforms RR for both utilities (R and

TABLE 7.2 Scheduling Results with Different Utility Functions: Four Users Case

Method	$\sum_{k=1}^{K} R_k$ (Mbps)	$\prod_{k=1}^{K} R_k$	$\sum_{k=1}^{K} \ln(R_k)$
Greedy	19.247	21.4178	58.3263
PFF	13.024	25.9986	58.5201
PFTF1	14.763	46.6128	59.1039
PFTF2	14.719	45.8719	59.0879
PFTF3	14.703	45.9718	59.0901

This table is reprinted, with permission from Elsevier, from E. Yaacoub and Z. Dawy, "Achieving the Nash Bargaining Solution in OFDMA Uplink Using Distributed Scheduling with Limited Feedback", *International Journal of Electronics and Communication AEU (Elsevier)*, 65(4), 320–330, 2011 ©2011 Elsevier.

TABLE 7.3 Scheduling Results with Different Utility Functions: Eight Users Case

Method	$\sum_{k=1}^{K} R_k$ (Mbps)	$\prod_{k=1}^{K} R_k$	$\sum_{k=1}^{K} \ln(R_k)$
Greedy	37.273	8.1008	112.6160
PFF	18.277	93.0743	115.0575
PFTF1	22.908	524.9444	116.7874
PFTF2	22.823	509.4447	116.7574
PFTF3	22.730	491.8786	116.7223

This table is reprinted, with permission from Elsevier, from E. Yaacoub and Z. Dawy, "Achieving the Nash Bargaining Solution in OFDMA Uplink Using Distributed Scheduling with Limited Feedback", *International Journal of Electronics and Communication AEU (Elsevier)*, 65(4), 320–330, 2011 ©2011 Elsevier.

TABLE 7.4 Scheduling Results with Different Utility Functions: 16 Users Case

Method	$\sum_{k=1}^{K} R_k$ (Mbps)	$\prod_{k=1}^{K} R_k$	$\sum_{k=1}^{K} \ln(R_k)$
Greedy	55.591	0	–
PFF	36.432	29.0824	224.4183
PFTF1	27.470	86.0293	225.5029
PFTF2	27.335	80.0188	225.4304
PFTF3	27.191	72.4726	225.3314

This table is reprinted, with permission from Elsevier, from E. Yaacoub and Z. Dawy, "Achieving the Nash Bargaining Solution in OFDMA Uplink Using Distributed Scheduling with Limited Feedback", *International Journal of Electronics and Communication AEU (Elsevier)*, 65(4), 320–330, 2011 ©2011 Elsevier.

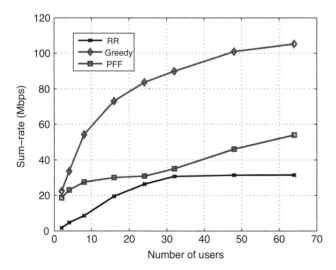

Figure 7.5 Sum-rate comparison of greedy and PFF utilities.

TABLE 7.5   User Distances from the BS

User	1	2	3	4	5	6	7	8
Distance (km)	0.2	0.3	0.4	0.5	0.6	0.7	0.85	1

$\ln(R)$) in terms of cell rate, and the greedy scheduling has a clear advantage over PFF. The RR algorithm allocates one RB for each user. This explains the linear increase in the sum-rate plot of RR until the number of users reaches the number of RBs. In this case, the number of served users during one TTI will be equal to the number of RBs, with one RB being allocated to each served user. The PFF plot in Fig. 7.5 confirms the theory of Theorem 7.1, since it shows that with PFF scheduling, the sum-rate increases and saturates as the number of users reaches the number of RBs. However, when the number of users exceeds the number of RBs, the sum-rate starts increasing since the scheduling behavior in this case results of allocating one RB to each of the $N_{\mathrm{RB}} < K$ users having the best channel conditions.

Fig. 7.5 does not show the effect of the distance from the BS on the rate of the different users. With greedy scheduling, users close to the BS are expected to receive most of the RBs, preventing edge users from fair access to resources most of the time. To obtain an indication about the fairness of the three investigated approaches, the case of eight users, located at fixed distances from the BS, is considered. The fixed distances considered are shown in Table 7.5.

Fig. 7.6 shows the rate results as a function of the distance. In the greedy scheduling scenario, users close to the BS achieve very high rate, whereas those close

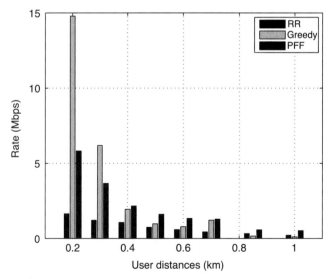

Figure 7.6   Rate achieved by each user as a function of the distance from the base station: 8 users and 25 RBs of 12 subcarriers each.

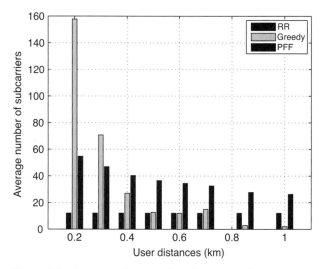

Figure 7.7   Average subcarrier allocation versus distance: 8 users and 25 RBs of 12 subcarriers each.

to the cell boundary suffer from starvation. The rate of the different users in the RR case is clearly very low regardless of the distance, but the rate gap between users in the center and the cell edge users is reduced. From Fig. 7.6 it can be seen that the PFF case allows cell edge users to achieve higher rate than RR and the greedy case, while allowing the users close to the center to keep the advantage of achieving a relatively higher rate than edge users.

In fact, studying fairness from the perspective of the average number of subcarriers allocated per TTI to each of the users of Fig. 7.6, the results of Fig. 7.7 are obtained. Naturally, in the RR case, all users are allocated the same number of subcarriers by definition. This achieves perfect fairness in terms of access to the resources for the different users. However, since RR is not a channel aware scheduling algorithm, this absolute fairness is not reflected into gains in the user rates, as can be seen from Fig. 7.6. On the other hand, with greedy scheduling, users close to the cell center are allocated most of the resources, leaving edge users with practically nothing, as can be seen for the user located at $d = 1$ km, where far more subcarriers are allocated to it by RR. However, with PFF scheduling, access to subcarriers is clearly more fair. All users, even at $d = 1$ km, are allocated more than one RB on average, as can be seen from Fig. 7.7, where the PFF outperforms RR even for the cell edge user (at $d = 1$ km).

7.5.4   Comparison to Existing Algorithms

The cell sum-rate results for the investigated algorithms with different utility functions are not displayed in figures since the plots of the different algorithms appear to overlap. Instead, the numerical results are listed in Table 7.6. It can be seen from

TABLE 7.6   Sum-Rate (Mbps) of the Different Algorithms

Number of Users	2	4	8	16	24	32	48	64
RR	1.72	4.70	8.53	19.44	26.31	30.67	31.36	31.41
Alg. 7.1–Greedy	22.41	33.72	54.41	73.26	83.79	90.04	101.06	105.37
Alg. 7.2–Greedy	22.35	33.62	54.25	73.11	83.67	89.90	100.93	105.20
Myung–Greedy	22.29	33.53	54.12	73.00	83.60	89.80	100.85	105.10
Ng–Greedy	22.40	33.69	54.32	73.16	83.71	89.98	100.97	105.22
Alg. 7.1–PFF	19.53	24.967	31.18	32.93	33.68	37.36	48.42	56.32
Alg. 7.2–PFF	18.62	22.98	27.46	30.00	30.87	35.06	46.11	53.95
Myung–PFF	18.21	22.47	26.87	29.97	30.13	35.24	46.30	54.15
Ng–PFF	18.21	23.25	28.86	31.81	33.54	37.33	48.41	56.31

Table 7.6 that in the greedy scheduling case, the four algorithms perform almost equally, which indicates that the algorithm with lowest complexity could be used at the expense of a negligible loss in performance. Clearly, Algorithm 7.1 has the best performance, followed by Ng's algorithm, then Algorithm 7.2, and finally Myung's algorithm. It should be noted that Algorithm 7.2 was able to outperform Myung's algorithm and achieve a performance within 0.3% of Algorithm 7.1 at a considerably lower complexity. In the PFF scheduling case, Algorithm 7.1 and Ng's algorithm have comparable performance, and they perform slightly better than Algorithm 7.2 and Myung's algorithm. Table 7.6 shows that Algorithm 7.1 outperforms Ng's algorithm and Algorithm 7.2 outperforms Myung's algorithm until the number of users increases beyond 32 where Myung's algorithm becomes slightly better. Nevertheless, Algorithm 7.2 achieves a performance within 5% of Algorithm 7.1 throughout the whole investigated range of the number of users.

To obtain an indication about the fairness of the investigated algorithms, the case of eight users located at fixed distances from the BS is considered, with the distances shown in Table 7.5. Fig. 7.8 shows the rate results as a function of the distance in the greedy scheduling, and Fig. 7.9 shows the average number of subcarriers allocated per TTI to each of the users of Fig. 7.8. The performance of the four algorithms is comparable. With all four algorithms, users close to the BS achieve very high rate, whereas those close to the cell boundary suffer from starvation.

Fig. 7.10 shows the rate results as a function of the distance in the PFF scheduling case, and Fig. 7.11 shows the average number of subcarriers allocated per TTI to each of the users of Fig. 7.10. In the PFF scenario, the superiority of Algorithm 7.1 is more evident. In fact, Fig. 7.10 shows that all users achieve a better rate with Algorithm 7.1, although Fig. 7.11 shows that users located beyond $d \geq 0.5$ km are allocated less resources by Algorithm 7.1 than by the three other algorithms. This indicates that Algorithm 7.1 allocates resources in a slightly more efficient way. Algorithm 7.2 performs better than Myung's algorithm and approximately as good as Ng's algorithm. Hence, Algorithm 7.2 can be used for linear complexity efficient scheduling in real-time applications with only a negligible degradation in performance.

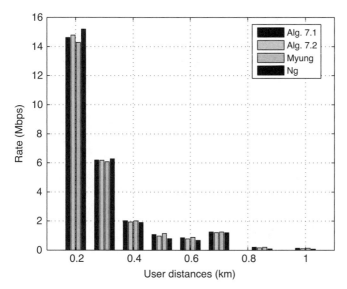

Figure 7.8 Rate achieved by each user as a function of the distance from the base station: 8 users and 25 RBs of 12 subcarriers each with greedy scheduling. This figure is reprinted, with permission from IEEE, from E. Yaacoub, H. Al-Asadi, and Z. Dawy, "Low Complexity Scheduling Algorithms for the LTE Uplink", IEEE Symposium on Computers and Communications (ISCC 2009), Sousse, Tunisia, July 2009 ©2009 IEEE.

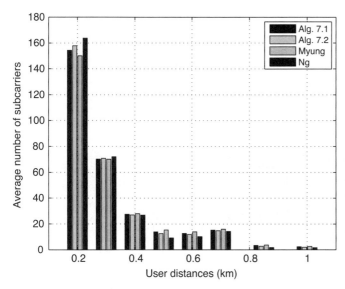

Figure 7.9 Average subcarrier allocation versus distance: 8 users and 25 RBs of 12 subcarriers each with greedy scheduling. This figure is reprinted, with permission from IEEE, from E. Yaacoub, H. Al-Asadi, and Z. Dawy, "Low Complexity Scheduling Algorithms for the LTE Uplink", IEEE Symposium on Computers and Communications (ISCC 2009), Sousse, Tunisia, July 2009 ©2009 IEEE.

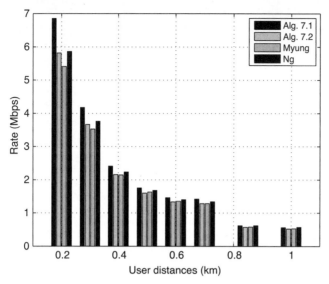

Figure 7.10 Rate achieved by each user as a function of the distance from the base station: 8 users and 25 RBs of 12 subcarriers each with PFF scheduling. This figure is reprinted, with permission from IEEE, from E. Yaacoub, H. Al-Asadi, and Z. Dawy, "Low Complexity Scheduling Algorithms for the LTE Uplink", IEEE Symposium on Computers and Communications (ISCC 2009), Sousse, Tunisia, July 2009 ©2009 IEEE.

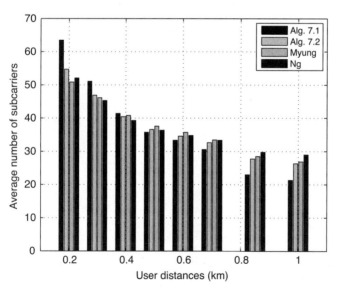

Figure 7.11 Average subcarrier allocation versus distance: 8 users and 25 RBs of 12 subcarriers each with PFF scheduling. This figure is reprinted, with permission from IEEE, from E. Yaacoub, H. Al-Asadi, and Z. Dawy, "Low Complexity Scheduling Algorithms for the LTE Uplink", IEEE Symposium on Computers and Communications (ISCC 2009), Sousse, Tunisia, July 2009 ©2009 IEEE.

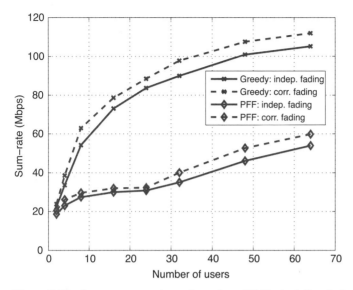

Figure 7.12   Sum-rate comparison of greedy and PFF scheduling: independent versus correlated fading.

7.5.5   Independent Versus Equal Fading over the Subcarriers of an RB

In Section 7.5.1, independent fading was assumed over each subcarrier. In this section, the case of IID fading per subcarrier is compared to the case of correlated fading, where equal fading is assumed over the subcarriers of a single RB and independent fading is assumed between RBs. The sum-rate results are shown in Fig. 7.12. It can be seen that a better performance is obtained with the correlated fading case. In the case of independent fading, all the subcarriers of a single RB are forced to be allocated together, although the fading might have different fluctuations on each subcarrier. However, in the case of equal fading, all subcarriers of a single RB have the same instantaneous fading value, and thus it is logical to allocate them together. It should be noted that, if subcarriers were allowed to be allocated individually, better performance would be reached with independent fading since frequency diversity would be used more efficiently in this case. However, in the worst case, the difference between independent and equal (maximum correlation) fading does not exceed 10%. In practice, the actual performance would be between these two extremes.

To investigate the utility results, Fig. 7.13 shows the plot of the PFF utility, that is, the logarithm of the rate. It should be noted that, in the simulations, when a user has a rate of zero, and to avoid dealing with $-\infty$ values, the rate of that user is set to a very low value ϵ instead, with $\epsilon = 2.2 \times 10^{-16}$ in Matlab, for example. This explains the linear decrease in the utility function $U = \ln(R)$ in Fig. 7.13 when K becomes greater than N_{RB}. Since no clear distinction can be made in Fig. 7.13, the results are

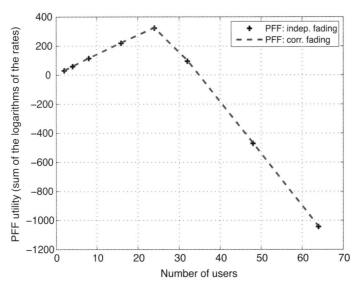

Figure 7.13   PFF utility comparison: independent versus correlated fading.

TABLE 7.7   PFF Utility

Number of users	2	4	8	16	24	32	48	64
Independent fading	29.6	57.8	113.6	219.2	322.7	94.3	−471.2	−1043.6
Correlated (equal) fading	30.1	59.7	115.8	222.2	325.9	100	−467.6	−1040.3

displayed in Table 7.7, where a slight superiority is shown for the case of correlated fading.

7.6   SUMMARY

Suboptimal scheduling in OFDMA uplink was extended to utility maximization with RB-based scheduling. Two algorithms with polynomial complexity were presented. Proportional fair scheduling was shown to be equivalent to the Nash bargaining solution maximizing the Nash product. Several utilities representing proportional fairness were investigated, and the proposed algorithms were shown to compare favorably with other algorithms in the literature.

This chapter closes the part of the book dedicated to centralized scheduling within a single cell. Chapter 8 represents an intermediate step between fully centralized and fully distributed single cell scheduling. Then, Chapters 9 and 10 deal with distributed scheduling with and without user cooperation, respectively.

SCHEDULING WITH DISTRIBUTED BASE STATIONS

Chapters 3–7 have dealt with centralized scheduling, where a central BS is responsible for the resource allocation within a cell. Chapters 9 and 10 treat the problem of distributed resource allocation, where mobile users take part in the scheduling process. This chapter[2] represents an intermediate step between centralized and distributed scheduling. In fact, in distributed scheduling scenarios, users should be in a relatively close proximity in order to exchange information or to "hear" the transmissions of other users. Thus, users in distributed scheduling should be within a reduced coverage area. Distributed base stations (DBSs) solve this problem, by dividing the coverage of a single cell into smaller coverage areas, with each area served by a remote radio head (RRH) connected to a central BS in the cell. Consequently, distributed scheduling could take place within the coverage area of a single remote radio head.

The chapter is organized as follows. Section 8.1 presents the necessary background and lists the main topics of this chapter. The system model is presented in Section 8.2. A scheduling algorithm suitable for a distributed base station scenario is described in Section 8.3. The simulation results are presented and discussed in Section 8.4. Discussions about the use of relays and femtocells instead of distributed base stations are presented in Sections 8.5 and 8.6, respectively. Finally, the chapter is summarized in Section 8.7.

8.1 BACKGROUND

The concept of distributed base stations and remote radio heads emerged to increase the coverage and capacity of wireless networks in a cost-effective way. It consists of a centrally located BS enclosure connected to RRHs via fiber optic cables [66]. In the existing literature, the terms distributed base station and distributed antenna system (DAS) are used interchangeably. DBSs were initially proposed to enhance

---

[2] This chapter is updated, with permission from John Wiley and Sons, from E. Yaacoub and Z. Dawy, "Uplink Scheduling in LTE Systems using Distributed Base Stations", *European Transactions on Telecommunications (ETT)*, 21(6), 532–543, 2010.

Resource Allocation in Uplink OFDMA Wireless Systems: Optimal Solutions and Practical Implementations, Elias E. Yaacoub and Zaher Dawy.

indoor coverage of cellular systems where a building is treated as a single cell with several distributed antennas rather than either multiple pico cells each with a dedicated antenna or as a single cell with one central antenna [67]. The DBS approach allows avoiding excessive handovers in the first case and significant fading in the latter. The coverage and capacity of DBSs in an indoor WCDMA system were investigated in Ref. [68] for several types of antennas. In a multicell CDMA system with DBSs, it was shown that maximum ratio combining (MRC) in the uplink achieves a considerable capacity and coverage enhancement, but simultaneous transmission in the downlink reduces performance since it increases the intercell interference [69]. A solution for this problem was proposed in Ref. [70], where it was found that selecting only the RRH with best channel to the user ensures the best downlink performance with DBSs. A similar conclusion was reached in Ref. [71] where transmitting from the RRH with the best channel was shown to outperform the case of using the RRH as a relay while transmitting the signal directly from the BS. These results were validated from an information theoretic standpoint in Refs [72, 73], where selective transmission (from only the RRH with best channel to user) was compared to maximum ratio transmission (using all the RRHs). In Ref. [122], it was shown that selection combining (SC) in the uplink provides considerable enhancement over centralized BSs and constitutes a good trade-off between performance and complexity when compared to MRC. A generalization of the concept of DBSs was presented in Ref. [123], where each RRH consists of several antennas ensuring microdiversity, and the set of RRHs contributes to macrodiversity.

It should be noted that, in a practical scenario, installing RRHs at desired locations (e.g., equidistant along the cell boundary) might not be possible. Therefore, the performance of random placement of RRHs throughout the cell was investigated in Refs [74, 75], in terms of outage probability, as a lower bound on the actual performance. Interestingly, it was found that as the number of RRHs increases, the performance converges to that of regularly deployed RRHs. In fact, in the case of both fixed and random RRH locations, the gains achieved by a DBS system were shown to increase with the number of RRHs up to a certain limit where the gain obtained after using an additional RRH is negligible. This limit was considered to be four and seven RRHs in Refs [70, 72], respectively, for the regular RRH positions, and seven in Ref. [75] for the random RRH positions. Hence, the investigated scenarios in this chapter will be selected within these limits. In Ref. [124], a solution is presented for DBS cooperation via block-diagonalization and dual-decomposition to maximize the weighted sum network capacity under per-antenna power constraint. The solution is a trade-off between intercell interference mitigation, spatial multiplexing, and macro diversity. In Ref. [125], a protocol to manage multiuser interference in a DBS system is proposed. The scheme is not specific to a particular multiple access scheme. It is assumed in Ref. [125] that users may connect to more than one RRH using the same resources, which leads to an increase in interference. The protocol of Ref. [125] deals with the problem of assigning channels to mobiles having dynamically changing sets of RRHs linking each of them to the network with a macrodiversity gain. Conversely to Refs [124, 125] where RRHs are considered to use the same resources without centralized control, the RRHs in this chapter are connected to a single BS, and the resources allocated to different RRHs are orthogonal subsets of the resources available at the central BS. Hence, contrarily to Refs [124, 125], interference is not an issue within a single cell.

Distributed BSs were implemented in a variety of wireless systems. In Ref. [126], they were proposed for LMDS. In Ref. [127], the commercial deployment of distributed BSs for WCDMA/HSPA was announced, and in Ref. [128] it was announced for cdma2000/EvDO. Although many of the DBS results in the existing literature apply to OFDMA, the impact of DBSs on user scheduling and resource allocation in OFDMA-based systems is not sufficiently investigated.

In this chapter, the problem of resource allocation in OFDMA systems with distributed BSs is treated in order to investigate the achievable performance gains. The main topics of this chapter can be summarized as follows:

1. Applying the concepts of DBSs in the context of OFDMA scheduling, and showing that considerable sum-rate and fairness enhancements can be achieved without any advanced techniques such as diversity with MRC through the antennas of the different RRHs, or selecting the optimal multiple input multiple output (MIMO) weights.

2. Investigating centralized scheduling using an algorithm initially proposed for centralized BSs in Chapter 7 and applying it in the case of DBSs.

3. Presenting an algorithm customized for scheduling in the case of DBSs, and showing that it has considerably less complexity than the centralized algorithm.

8.2 SYSTEM MODEL

The model is composed of a single central BS connected to several RRHs distributed throughout the cell area. The BS controlling the RRHs could be colocated with any of the RRHs or in a separate location. Although other types of media are possible, it is mainly connected to RRHs via fiber optic cable. Connection topologies include star, chain, tree, and ring topologies [66]. Each RRH consists mainly of a remote antenna connected to the central BS. This allows centralized control to be performed by the BS as in the conventional case while the RRHs allow extended coverage and/or more user capacity. In addition, for fixed coverage and user capacity, the RRHs provide the users with better quality of service (QoS) since the distance from a user to the nearest RRH will be smaller than the distance to the central BS antenna in the conventional case, which leads to a higher SNR. Fig. 8.1 shows examples of distributed BS deployment scenarios. In Fig. 8.2, an example of six users connected to the five RRHs of deployment scenario (c) is shown.

In this book, distributed BSs will not be investigated in terms of increasing the cell coverage and/or capacity. A single cell scenario is considered, and the performance of the studied scheduling algorithms is compared in terms of sum-rate and fairness. In the comparisons, the same coverage area and the same number of users are considered in the cell in the case of a centralized BS and distributed BS. The same number of subcarriers is also considered in both cases. Hence, the RRHs are not used to ensure more frequency reuse, but rather to make a more efficient and fair use of the available subcarriers.

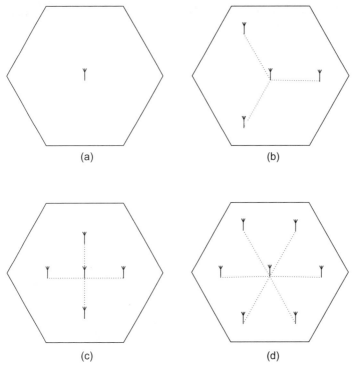

Figure 8.1 Examples of possible deployment scenarios.

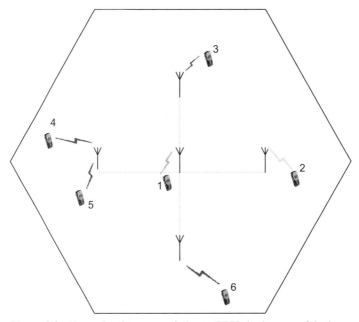

Figure 8.2 Example of user association to RRHs in the case of deployment scenario (c).

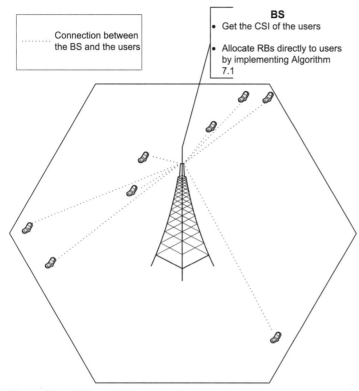

Figure 8.3 CBS scheduling scenario.

The investigated and compared scenarios consist of the following:

- *Centralized BS and Centralized Scheduling (CBS)*: A single central BS with an omnidirectional antenna covering the whole cell. A centralized scheduling algorithm (Algorithm 7.1 presented in Chapter 7) is run at the central BS location. This scenario is illustrated in Fig. 8.3.

- *Distributed BS and Centralized Scheduling (DBS-C)*: A central BS with several RRHs throughout the cell. A centralized scheduling algorithm (Algorithm 7.1 presented in Chapter 7) is run at the central BS location to allocate resources to the users. The presence of the RRHs contributes in enhancing the channel states of the different users by providing each user with an antenna that is closer to it than the central BS antenna. Hence, the scheduling operation is similar to the previous scenario but with better CSI available at the BS for the different users. This scenario is illustrated in Fig. 8.4.

- *Distributed BS and Distributed Scheduling (DBS-D)*: A central BS with several RRHs throughout the cell. A scheduling algorithm for DBS scenarios (Algorithm 8.1 presented in Section 8.3) is run at the locations of the RRHs to allocate resources to the users. The presence of the RRHs contributes in enhancing the channel states of the different users as in the previous case (DBS-C). Furthermore, the complexity of implementing the scheduling algorithm at the various

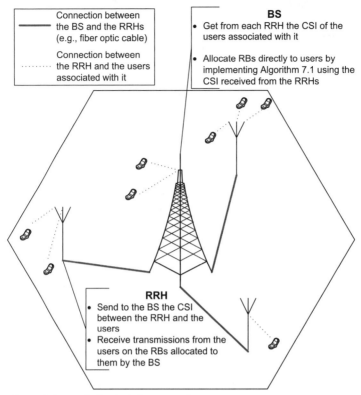

Figure 8.4 DBS-C scheduling scenario.

locations is less than implementing it at the central location for the whole cell. This scenario is illustrated in Fig. 8.5. It should be noted that in the (DBS-D) case, the scheduler for each RRH does not have to be colocated with the corresponding RRH. A single scheduler in the central BS location could be used to perform scheduling operations separately for each RRH and communicate the scheduling information via the medium connecting the BS to the RRHs (e.g., dedicated fiber optic cables).

8.3 SCHEDULING WITH DISTRIBUTED BASE STATIONS

In this section, a scheduling algorithm is presented for distributed BS scenarios. The algorithm is based on Algorithm 7.1 developed in Chapter 7. The DBS scheduling algorithm has considerably less complexity than Algorithm 7.1. The performance of the two algorithms is compared in Section 8.4.

8.3.1 Scheduling Algorithm for DBS Scenarios

The presented DBS scheduling algorithm, Algorithm 8.1, consists of two allocation phases. In the first phase, RBs are allocated to RRHs, with the amount of RBs allocated

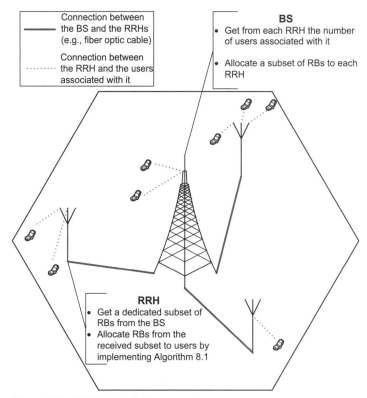

Figure 8.5   DBS-D scheduling scenario.

to each RRH being proportional to the number of users associated with this RRH (This association will be discussed in Section 8.4.5). In the second phase, Algorithm 7.1 is applied for each RRH separately.

Algorithm 8.1

> **Scheduling Algorithm for DBS Scenarios**
> **After the users are connected to the network, each subset of users is associated to a certain RRH, with user k associated to RRH a if the association would lead to the best SNR for user k. It is assumed that the users have low mobility such that they remain associated to the same RRH throughout the transmission time.**
> **Step 1: Allocate $N_a = \lfloor N_{RB} K_a / K \rfloor$ RBs to each RRH a with K_a users connected to it.**
> **Step 2: Due to the floor operation, some RBs will remain unallocated. Allocate sequentially one RB per RRH until all RBs are allocated.**
> **Step 3: Apply Algorithm   7.1 sequentially to each RRH; that is, for RRH a, apply Algorithm   7.1 with $K = K_a$ and $N = N_a$.**

8.3.2 Complexity Analysis of the DBS Scheduling Algorithm

It was shown in Chapter 7 that Algorithm 7.1 has approximately $\mathcal{O}(N_{RB}^2 K)$ complexity. As for Algorithm 8.1, the first phase has a complexity of the order of the number of RRHs. Since in most cases (and particularly the ones investigated here), the number of RRHs is negligible compared to $N_{RB}^2 K$, it is the complexity of the second phase that will dominate. With uniform distribution of the users inside the cell, and assuming the RRHs are distributed such that they cover equal areas in the cell, it is logical to assume that, as the number of users increases, the following expressions are satisfied: $N_a = N_{RB}/A$ and $K_a = K/A$, with A being the number of RRHs. Hence, the complexity of the second phase of Algorithm 8.1 is given by

$$\mathcal{O}\left(\sum_{a=1}^{A} N_a^2 K_a\right) \approx \mathcal{O}\left(\sum_{a=1}^{A} \frac{N_{RB}^2 K}{A^3}\right)$$
$$= \mathcal{O}\left(\frac{N_{RB}^2 K}{A^2}\right) \tag{8.1}$$

Consequently, the complexity of the DBS scheduling algorithm decreases with the square of the number of RRHs. This result is interesting, since it shows that the complexity can be reduced by around an order of magnitude with only three RRHs. It should be noted that in case Algorithm 7.2 is applied instead of Algorithm 7.1 in Step 3 of Algorithm 8.1, the complexity would be

$$\mathcal{O}\left(\sum_{a=1}^{A} N_a K_a\right) \approx \mathcal{O}\left(\sum_{a=1}^{A} \frac{N_{RB} K}{A^2}\right)$$
$$= \mathcal{O}\left(\frac{N_{RB} K}{A}\right) \tag{8.2}$$

Hence, with Algorithm 7.2, the complexity of the DBS scheduling algorithm decreases with the number of RRHs. This also constitutes an important reduction in complexity, since Algorithm 7.2 has a linear complexity in the number of RBs compared to a quadratic complexity for Algorithm 7.1.

8.4 RESULTS AND DISCUSSION

This section presents the simulation results obtained by applying Algorithms 7.1 and 8.1 in a DBS scenario. The results include plots of the cell sum-rate, in addition to a discussion of fairness.

8.4.1 Simulation Model

The simulation model consists of a single cell with a BS equipped with an omnidirectional antenna, or consisting of several RRHs, each consisting of an omnidirectional antenna. The investigated RRH deployment models are shown in Fig. 8.1.

Deployment scenario (a) consists of the conventional centralized single BS. Deployment scenario (b) consists of four RRHs: one located at the cell center and three located at a distance of $2R_c/3$, with R_c being the cell radius considered to be 1 km. The angular separation between these three RRHs is $120°$. Deployment scenario (c) consists of five RRHs: one located at the cell center and four located at a distance of $R_c/2$, with $90°$ angular separation between them. Finally, deployment scenario (d) consists of seven RRHs: one located at the cell center and six located at a distance of $2R_c/3$, with $60°$ angular separation between them. In each deployment scenario, the performance of the three scheduling cases is investigated: CBS, DBS-C, and DBS-D.

The rate is averaged over 100 TTIs, with the duration of a TTI being 1 ms. Then the simulation is repeated over 100 iterations. The total bandwidth considered is $B = 5$ MHz, subdivided into 25 RBs of 12 subcarriers each [41]. A target BER of 10^{-6} is considered. The maximum mobile transmit power is considered to be 125 mW [20]. All mobiles are assumed to transmit at the maximum power, and the power is subdivided equally among all subcarriers allocated to the mobile. The channel gain over subcarrier i corresponding to user k is given by (7.15). Rate calculations are performed as in Section 7.3.3, but with d_k denoting the distance in km from mobile k to the nearest RRH in the distributed BS scenario, instead of the distance to the BS in the centralized BS scenario of Chapter 7.

8.4.2 Sum-Rate Results

Throughout this section, the results corresponding to round robin will be denoted by "RR", and those corresponding to the presented algorithms with greedy and PF scheduling will be denoted by "Greedy" and "PFF", respectively. Users are distributed uniformly within the cell area. Figs 8.6–8.8 compare the results of the

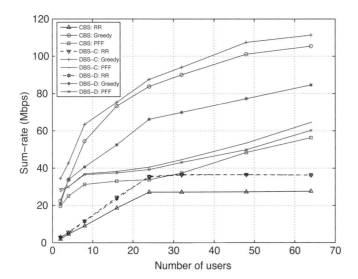

Figure 8.6 Sum-rate comparison in the case of deployment scenario (b).

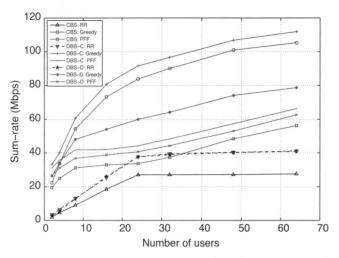

Figure 8.7   Sum-rate comparison in the case of deployment scenario (c).

different algorithms in the case of deployment (a), to deployments (b), (c), and (d), respectively.

These figures show that for greedy scheduling, the case DBS-C outperforms the case CBS, which in turn outperforms the case DBS-D. This is due to the allocation of RBs to each RRH proportionally to the number of users in Algorithm 8.1 (case DBS-D), which hinders the concept of greedy scheduling that would allocate most RBs to users nearest to RRHs (case DBS-C), not proportionally to each RRH.

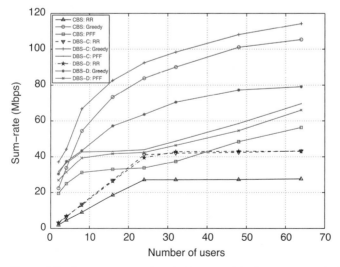

Figure 8.8   Sum-rate comparison in the case of deployment scenario (d).

For PFF, the case DBS-C outperforms the case DBS-D, which in turn outperforms the case CBS. It should be noted that the performance of DBS-C and DBS-D scheduling is the same with RR, since each user is allocated a single RB in both cases. In addition, these figures show the expected result that in terms of rate, greedy scheduling outperforms PF scheduling, which in turn outperforms RR scheduling. However, it should be noted that Figs 8.6–8.8 show the sum-rate, but do not show the effect of the distance from the BS or RRH on the rate of the different users. With greedy scheduling, users close to the BS or RRH are expected to receive most of the RBs, preventing edge users from fair access to resources most of the time. These issues will be discussed in the following section.

Comparing Figs 8.6–8.8, it can be seen that as the number of RRHs increases, the performance of RR increases. Although the number of RBs allocated to each user is the same, the sum-rate increases since each user's chances of finding a nearby RRH increase with the number of RRHs. This leads to an increase in SNR and thus in the achieved rate. In addition, with greedy scheduling, the superiority of the scenarios with more RRHs decreases as the number of users increases. In fact, with a high number of users, the probability of some users being near an RRH and receiving most of the resources increases, even with a low number of RRHs. With PFF, the performance enhancement increases with the number of RRHs for both Algorithms 7.1 and 8.1.

8.4.3 Fairness Analysis

To obtain an indication about the fairness of the different investigated algorithms in different cases, the example of Fig. 8.2, which consists of six users located at fixed positions in the case of scenario (c), is considered. Their positions is expressed in polar coordinates, that is, a distance and an angular position from the origin, taken to be the cell center. The coordinates of the six users are shown in Table 8.1.

Fig. 8.9 shows the average rate results of each user in the greedy scheduling case, and Fig. 8.10 shows the average number of subcarriers allocated per TTI to each of the users of Fig. 8.9. With the CBS case and greedy scheduling, almost all resources are allocated to user 1, the closest to the BS, which achieves the highest rate. User 5, the second nearest user, receives a very small amount of resources, whereas all other users suffer from starvation. In the DBS scenarios, the resource allocation process is clearly more fair. Cases DBS-C and DBS-D with greedy scheduling allow all users except user 1 to receive more resources and achieve considerably higher rate than the centralized case (CBS).

Fig. 8.11 shows the average rate results of each user in the PFF case, and Fig. 8.12 shows the average number of subcarriers allocated per TTI to each of the users of Fig. 8.11.

TABLE 8.1 User Positions with Respect to the Cell Center

User	1	2	3	4	5	6
Distance (km)	0.15	0.9	0.85	0.95	0.5	1
Angle (°)	190	0	80	160	200	290

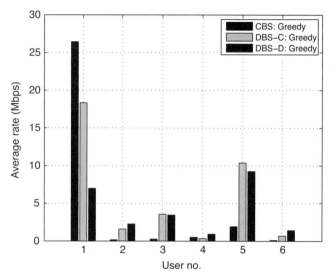

Figure 8.9   Average rate achieved by each user: greedy scheduling.

With PFF scheduling, Cases DBS-C and DBS-D clearly outperform the CBS case in terms of achieved rate and subcarriers allocated, except for user 1. Contrarily to the greedy scheduling case, the performance of cases DBS-C and DBS-D with PFF is comparable, with a slight superiority for DBS-C. In addition, PFF scheduling is clearly more fair than with greedy scheduling. The most striking example can be

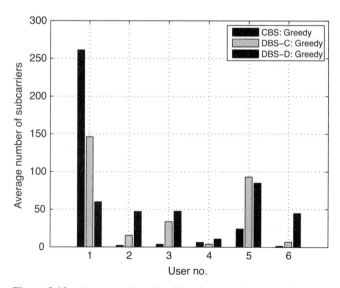

Figure 8.10   Average subcarrier allocation: greedy scheduling.

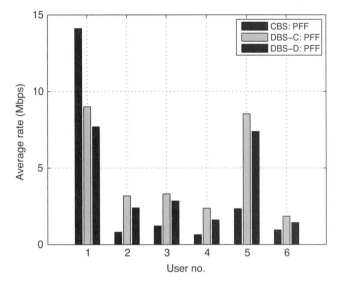

Figure 8.11   Average rate achieved by each user: PFF scheduling.

seen in the case of users 4 and 5, which are associated with the same RRH. Figs 8.9 and 8.10 show that cases DBS-C and DBS-D with greedy scheduling clearly favor user 5, the nearest to the RRH, whereas Figs 8.11 and 8.12 show that cases DBS-C and DBS-D with PFF scheduling are considerably more fair toward user 4.

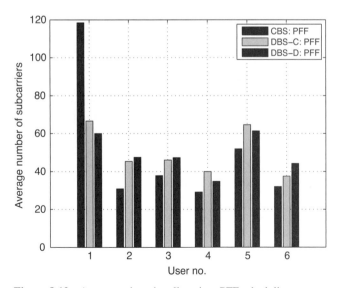

Figure 8.12   Average subcarrier allocation: PFF scheduling.

8.4.4   Location Optimization

In this section, the effect of optimizing the locations of the RRHs on the overall cell rate is investigated. Scenario (d) is considered as an example. In Section 8.4.2, deployment scenario (d) was investigated with one RRH located at the cell center and six RRHs located at a distance of $2R_c/3$, with $60°$ angular separation between them. In this section, the RRHs are placed such that the areas covered by each RRH are equal. In fact, with a uniform user distribution throughout the cell and with the RRHs having the same characteristics (transmit power, antenna, etc.), dividing the load equally among the RRHs is the best solution. Hence, fixing one RRH at the cell center, the cell is divided into seven equal coverage areas and one RRH is placed at the centroid of each area, as shown in Fig. 8.13. The results are compared to those of scenario (d) in Section 8.4.2, for both the DBS-C and DBS-D cases, with utilities $U = R$ and $U = \ln(R)$. The results are displayed in Fig. 8.14. Although the optimized positions yield better results as expected, the enhancement is less than 5% between the optimized distance and the positions selected in Section 8.4.2. Furthermore, the enhancement increases with the number of users with greedy scheduling, but it is negligible with PFF scheduling. Hence, in a practical deployment scenario, deviating slightly from the optimized positions does not lead to a noticeable degradation in performance. This agrees with the results of Ref. [75] in that the position of the RRHs does not severely affect the performance, as long as the number of RRHs is sufficient.

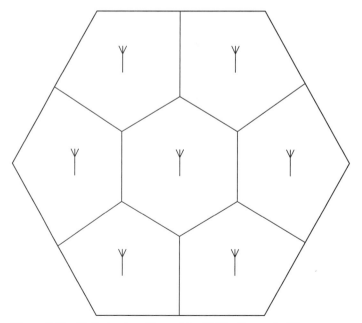

Figure 8.13   Optimized positions of the RRHs of scenario (d).

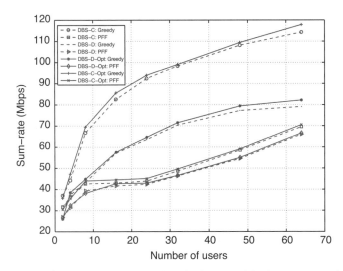

Figure 8.14   Sum-rate comparison in the case of deployment scenario (d): optimized versus nonoptimized RRH locations.

8.4.5 Mobility Considerations

In the simulations of the previous sections, the mobile users were considered with limited or no mobility. It would seem logical to expect that in higher mobility scenarios, handover rates between the different RRHs and the corresponding signaling overhead would reduce the performance of the system. However, this is not the case. In the considered model, the presence of the RRHs is transparent to the users who act as if there was only a single central BS in the cell. The signaling and allocation of resources to RRHs take place at the central BS location. Below is a proposed approach for this allocation.

- At the beginning, the BS allocates the RBs equally among all RRHs, with an equal separation between RBs to ensure frequency diversity.
- The BS extracts the necessary CSI from the SRS transmitted by the mobile in order to perform channel dependent scheduling [98].
- From the measured CSI, if the estimated SNR of a given user k averaged over the subcarriers allocated to RRH j is higher than the SNR averaged over the subcarriers allocated to other RRHs, then user k is associated to RRH j.
- After determining the number of users linked to each RRH, the BS allocates the RBs to RRHs appropriately (e.g., proportionally to the number of users in case DBS-D). Naturally, as the number of users increases, the uniform user distribution assumption will eventually lead to a nearly equal allocation of subcarriers in the case of optimal RRH positions.
- The association of users to RRHs can be modified depending on the changes in the SRS measurements.

Hence, the distributed BS system appears as a single BS to the user. The RRH acts as a simple BS antenna located closer to the user position. The communication between the central BS and the RRHs is via fiber optic (or microwave links having a nonoverlapping spectrum with the wireless OFDMA network) and consequently does not consume any OFDMA radio resources. The mobility of a user heading from the area of RRH A to the area covered by RRH B is translated in terms of decreasing channel quality on the subcarriers of RRH A and better channel quality on the subcarriers of RRH B. Hence, the user is allocated a subset of the subcarriers of RRH B similarly to what would happen in a single BS scenario when channel quality deteriorates on certain subcarriers and improves on others. The presence of RRHs and the actual handover occurrence are oblivious to the user. As for mobility between cells, the handover rules apply as in the case of two cells with a central BS in each.

8.5 DISTRIBUTED BASE STATIONS VERSUS RELAYS

Increasing both the capacity and the coverage of traditional cellular architectures requires the deployment of a large number of BSs, which leads to excessive costs for network service providers [129, 130]. The approach of DBSs and RRHs studied in this chapter is an efficient method to deal with this problem. However, a wired connection between the central BS and the RRHs is not always feasible or easily realizable in practice. In this case, this wired connection is substituted by a wireless connection, and relay stations (RSs) are used instead of RRHs. Thus, the RS with less functionality than the BS can forward the data to remote areas of the cell and/or increase the cell capacity with reduced infrastructure cost. An example is shown in Fig. 8.15.

In relay enhanced wireless networks, resource allocation imposes an additional challenge than scheduling with RRHs. In fact, with relays, resource allocation should be performed on the links between the BS and RS, in addition to the links between the BS/RS and the mobile users. Thus, efficient scheduling decisions should be coupled with routing methods in order to ensure the best possible management and allocation of the wireless resources. In fact, various channel gains are generally achieved on the various links along each route in a communication system with RSs. Consequently, the routing schemes adopted affect the performance results reached [131]. Joint routing and resource allocation is known to lead to better results compared to separate scheduling and routing in the presence of RSs [132].

Hence, resource allocation in OFDMA relay-based networks is gaining increasing research interest and OFDMA-based relaying is included in state-of-the-art and next generation wireless communications systems [90, 94]. In Refs [133–137] centralized resource allocation is performed in single cell OFDMA networks with RSs. CSI knowledge is assumed at the BS. An OFDMA single cell network with a single RS is studied in Ref. [133], whereas a network with multiple RSs is studied in Ref. [134]. In Refs [133, 134], fixed RSs are assumed. In Ref. [133], an SNR-based path selection algorithm is executed to determine the node (BS or RS) to route the data of a mobile user, and scheduling is decoupled from routing. CSI feedback from the mobile to the BS or RS is performed every time frame. In Ref. [134], the objective

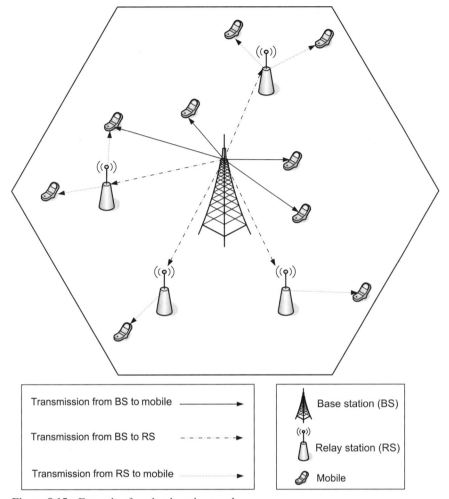

Figure 8.15 Example of a relay-based network.

function considered is the total average throughput on both the direct and relayed links. In Ref. [135], an OFDMA network with multiple RSs and a single destination is studied. Load balancing constraints are imposed on the RSs. Furthermore, it is assumed that the subcarriers allocated in the first hop are the same as those allocated in the second hop, which leads to a reduced benefit from the frequency diversity gain. In Refs [136, 137], a single cell OFDMA system is considered, with the BS surrounded by three RSs. Fixed subcarriers are allocated to the BS–RS links, and resource allocation is performed on the links from the BS or RSs to the mobile users. The approach of Refs [136, 137] performs joint routing and scheduling, by resorting to adaptive subcarrier, path, and power allocation to maximize the system capacity.

Centralized resource allocation in multicell OFDMA relay-based networks is investigated in Refs [138, 139]. In Ref. [138], a centralized scenario in a multicell OFDMA network with six fixed RSs per cell is considered. In the approach of Ref. [138], relay selection is based on the maximum power received, without considering the signal to interference and noise ratio (SINR). Furthermore, only log-normal shadowing is considered (without Rayleigh fading). The investigated problem in Ref. [138] consists of determining the minimum number of subcarriers to satisfy the QoS requirements of a given user. Thus, no techniques are used for mitigating cochannel interference. In Ref. [139], routing is decoupled from resource allocation. Mobile users are grouped into two categories: near users that communicate directly with the BS, and far users that can only receive data in two-hop transmissions through the closest RS. Two different allocation policies are considered in Ref. [139]. The first policy enforces the constraint that there is no frequency reuse on the following three links: the BS–RS link, the RS-far user link, and the BS-near user link. These links are activated for only a fraction of the total time. In the second policy, the BS–RS link is active within a fraction of the total transmission time, while the BS-near user and RS-far user links are active simultaneously during the remaining time portion. Thus, the latter concurrent transmissions could lead to intracell interference, since no interference coordination is adopted between the BS and the RSs in Ref. [139]. However, a relay reuse factor is adopted such that a subcarrier on the RS-far user link can be reused in the same cell. It is shown in Ref. [139] that capacity enhancement is obtained due to the spectrum reuse among the BS and RSs, in addition to multiuser diversity gains.

Centralized allocation schemes necessitate a significant signaling overhead. With RSs, this overhead increases, since the BS needs CSI information about the BS–RS links in addition to the BS–user and RS–user links. Furthermore, resource allocation decisions should be communicated to the RSs and users. Hence, the investigation of distributed resource allocation schemes with RSs is in order, although the contributions in this area have been limited so far [132].

In Ref. [140], a semidistributed scheme in a single cell OFDMA system with half-duplex fixed relays is considered. The BS allocates resources directly to the BS-near user links, and allocates resources to the RS-far user links through the RSs. It is implicitly assumed in Ref. [140] that routes have been established before resource allocation, regardless of the channel conditions, and that the same subcarrier is used on the two hops, BS–RS and RS–user. Furthermore, a protocol for gathering CSI is assumed to be available. In addition, the allocation decisions are considered to be broadcasted on a separate control channel.

The approach of Ref. [140] follows two steps. In the first step, an RS and the users associated to it are considered as a virtual single user by the BS, which allocates its resources to near-users as well as the virtual users. The virtual user is represented by the RS, and the rate requirement of each virtual user is considered to be the sum of the rate requirements of the users associated with that RS. In the second step, each RS allocates resources to the users in its coverage area. Two allocation methods are considered. In the first method, resources allocated in the first step to the BS–RS link are distributed among the users assigned to that RS. This method has comparable performance to the centralized case. In the second method, each RS allocates all the available subcarriers to the users within its coverage area, which leads to intracell interference. Both schemes are shown to reduce the amount of overhead required to feedback the CSI and minimum rates to the BS. However, these schemes decouple routing and resource allocation and do not exploit the interference avoidance and traffic diversity gains.

Hence, although OFDMA-based relays are part of the state-of-the-art and next generation wireless communications systems [90, 94], optimizing joint routing and resource allocation in OFDMA networks with relays still represents an interesting research challenge.

8.6 DISTRIBUTED BASE STATIONS VERSUS FEMTOCELLS

The demand for pervasive wireless access and the need for high data rates are expected to grow significantly in the foreseeable future with the proliferation of novel resource demanding applications such as gaming, mobile TV, and social networking. This motivates the introduction of new cost-effective approaches for expanding existing wireless networks efficiently. A recent solution in this direction is the introduction of femtocells that are low cost, low power, access points that can be overlaid with an existing wireless network [141–145]. These femtocells are expected to be small and inexpensive plug-and-play devices that can be installed both by the service providers in their network and by the end users at home. The femtocells are devices that can coexist with existing wireless infrastructure and are interconnected by an IP backhaul network through a local broadband connection, such as cable or digital subscriber line (DSL). In next-generation wireless networks, femtocells are expected to be ubiquitous, as home-users can buy their own access points at a low cost, hence, setting up their own personal cell.

The use of femtocells is expected to improve the overall performance of the wireless users by offloading traffic from the microcell/macrocell network through dedicated capacities provided to resource demanding locations such as homes or enterprises. This can lead to reducing the number of BS sites required to cover a given area, and hence reducing costs for network operators. An example is shown in Fig. 8.16 to demonstrate the interplay between femtocell deployments on one hand and BS site deployment and resource allocation on the other hand. It shows that the number of BS sites can be reduced when indoor traffic offloading via femtocells is taken into account, similarly to the case where RRHs are deployed in the cell. However, RRHs in a DBS scenario are connected the central BS, which has centralized control over the resource allocation process. Femtocells are independent devices, and there is no central entity that has control over femtocells in a given area. Hence, neighboring femtocells can reuse the same subcarriers, conversely to the approach followed in this chapter, where the sets of subcarriers allocated to different RRHs within a given cell are exclusive. In the general case, the same subcarriers can be allocated to the RRHs associated to a single DBS. This mandates the use of interference management techniques to mitigate the interference between the RRHs. This scenario is discussed in Chapter 11, Section 11.4. As for interference mitigation in the case of femtocells, the techniques discussed in Chapter 12 can be applied.

Hence, femtocells can potentially provide high user capacities while reducing the overall transmitted power, by bringing the users and their access points closer to

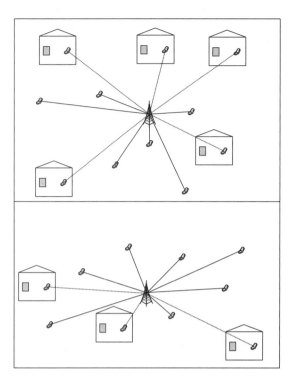

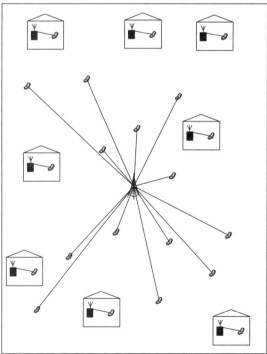

Figure 8.16 Impact of femtocell deployment on resource allocation. Upper part: resource allocation without the presence of femtocells; all users are connected to macrocell BSs, and two BS sites are needed for efficient resource allocation. Lower part: resource allocation with the presence femtocells; less users are connected to macrocell BSs, and one BS site is needed for efficient resource allocation.

one another. Nonetheless, the deployment of femtocells yields numerous technical challenges, notably due to the need for coexistence between the femtocells and the main wireless network. These challenges are discussed in Chapter 12, Section 12.6.4.

8.7 SUMMARY

The concept of distributed base stations was investigated in the context of OFDMA uplink. Resource allocation with a distributed base station was compared to resource allocation with a centralized base station. In the distributed base station case, both centralized and DBS scheduling algorithms were compared. Utility functions used in the scheduling algorithms included the rate and the logarithm of the rate. Monte Carlo simulations showed that centralized scheduling with a distributed base station achieves more rate and fairness than centralized scheduling with a centralized base station. In addition, in the distributed base station case, it was shown that the DBS scheduling algorithm achieves almost the same results as centralized scheduling with a proportional fair utility, with considerably reduced complexity. Discussions about the use of relays and femtocells were presented, and the differences with distributed base stations were outlined.

In the following two chapters, distributed resource allocation, where mobile users have a significant role in the resource allocation process, will be investigated in detail. Chapter 9 deals with distributed resource allocation with user cooperation, whereas Chapter 10 studies the noncooperative scenario.

DISTRIBUTED SCHEDULING WITH USER COOPERATION

In Chapter 8, distributed base stations were investigated in the context of OFDMA uplink scheduling. With DBSs, the cell coverage area is divided into smaller coverage areas, with each area served by an RRH connected to the central BS in the cell. Consequently, distributed scheduling could take place within the coverage area of a single RRH, since in distributed scheduling scenarios, users should be in a relatively close proximity in order to exchange information or to "hear" the transmissions of other users.

In this chapter, distributed scheduling with user cooperation is investigated, whereas Chapter 10 treats the noncooperative scenario. This chapter is organized as follows. Section 9.1 presents the background information and lists the main topics investigated in the chapter. The collaborative scheduling approach is described in Section 9.2. A collaborative distributed scheduling algorithm inspired from Algorithm 7.2 is presented in Section 9.3. The simulation results are presented and discussed in Section 9.4, and the chapter is summarized in Section 9.5.

9.1 BACKGROUND

Conversely to centralized resource allocation, mobile users have more autonomy in making transmission decisions in distributed schemes. The benefits of distributed resource allocation are being widely investigated in the context of ad-hoc networks, relay-based networks, and sensor networks [53–55], in addition to wireless local area networks (WLANs) [56, 57] and cognitive radio (CR) networks [33, 58–63].

In this chapter, a distributed uplink scheduling scheme for OFDMA systems is investigated. The studied scheme is based on the collaboration of mobile users. Collaboration consists of exchanging CSI between the users. To limit the overhead due to this exchange, a CSI quantization method is presented in order to reduce the number of required feedback bits. the presented scheme leads to results close to the centralized case, especially with PF scheduling. The PF solution is desirable in distributed scenarios since it is equivalent to the NBS, and hence it is in the benefit

Resource Allocation in Uplink OFDMA Wireless Systems: Optimal Solutions and Practical Implementations,
Elias E. Yaacoub and Zaher Dawy.
© 2012 by the Institute of Electrical and Electronics Engineers, Inc. Published 2012 by John Wiley & Sons, Inc.

135

of the users to cooperate in order to implement it, since it corresponds to their Pareto efficient solution from a game theoretical perspective.

The main topics discussed in this chapter are summarized as follows:

1. Investigating collaborative distributed scheduling in OFDMA uplink and presenting a distributed scheduling scheme that can be easily implemented by the users with a limited exchange of CSI.

2. Presenting a CSI quantization technique in order to achieve a near optimal performance with one bit feedback per subcarrier.

3. Presenting a low complexity scheduling algorithm by extending Algorithm 7.2 proposed in Chapter 7 for the full CSI case so that it can be implemented by the users in a distributed collaborative scenario with limited feedback.

9.2 COOPERATIVE DISTRIBUTED SCHEDULING SCHEME

In distributed scheduling, users have more independence since they take part in the scheduling decisions. In such a scenario, the BS would allocate a subset of the subcarriers to a group of users that are close enough to communicate with each other. In a game theoretical formulation, the users would then perform a bargaining game to share the allocated subcarriers, with each user trying to maximize its share. Obviously, the solution to this problem is the NBS, which is the optimal (Pareto efficient) solution. The challenge in such a scenario is to determine an efficient low-overhead communication protocol between the users that allows them to exchange the information necessary to perform the scheduling operation (e.g., broadcast of channel state information of each user on each subcarrier). Although greedy and PF scheduling will be investigated, more emphasis will be given to PF scheduling since it is more justified in a distributed framework, due to its equivalence to the NBS as discussed in Section 7.2.2.

9.2.1 System Model

The system studied consists of a single central controlling device (CCD) covering an area of limited size. The reduced coverage area is assumed because the users are required to be in a relatively close proximity so that they can collaborate by exchanging information. In this book, the term CCD is used in a distributed scheduling scenario in order to represent an entity taking the role of the BS in a centralized scenario or the RRH in a DBS scenario. The introduction of this new term is justified since it encompasses a broader concept than a BS or RRH. In fact, a CCD can represent in practice: a BS serving a small coverage area, a remote antenna or RRH in a distributed BS system, an access point (AP) in a local area network, a central controller in a CR network, or a femto BS inside a house or building. In this chapter, the performance of the presented scheme is investigated within the range of a single CCD due to

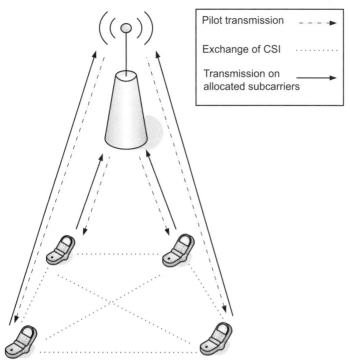

Figure 9.1   System model of the distributed collaborative scheduling scenario. This figure is reprinted, with permission from Elsevier, from E. Yaacoub and Z. Dawy, "Achieving the Nash Bargaining Solution in OFDMA Uplink Using Distributed Scheduling with Limited Feedback", *International Journal of Electronics and Communication AEU (Elsevier)*, 65(4), 320–330, 2011 ©2011 Elsevier.

the orthogonality of the subcarrier allocations. The model is shown in Fig. 9.1. The scheduling process takes place as follows:

- The CCD indicates to the users the available subcarriers for uplink transmission, by transmitting appropriate pilot signals on these subcarriers. In a cellular system with full reuse, the users are generally aware of these subcarriers and use the pilot signals for channel estimation. In other cases, for example, in a CR network, the CCD indicates to the users the free subcarriers available for transmission in order to avoid interference to primary users.

- Each user estimates its CSI on each of the available subcarriers. The CSI is estimated on the channel between the CCD and user. In a time division duplex (TDD) system, the users can determine their uplink CSI from downlink signals. In a frequency division duplex (FDD) system, the subcarriers used for the uplink are different from those used for the downlink. In this case, each uplink TTI should start with a pilot transmission phase where the CCD sends pilot signals on the subcarriers used for the uplink. This will not interfere with uplink transmission since the users are forbidden from transmitting during this phase.

This method can be used to synchronize the users with the CCD in an FDD system, since users have to wait for the pilot transmission phase in order to start the uplink scheduling process.

- After calculating the CSI on each subcarrier, each user broadcasts its CSI so that the other users can use it while implementing the scheduling algorithm. Several techniques can be used to perform the CSI broadcast: the broadcast can take place over dedicated subcarriers not included in the scheduling process, one for each user. Another approach would be to encode the CSI information using CDMA and transmit it over a single subcarrier dedicated for the exchange of CSI by all users. Another solution would be to use a completely different system, for example, Bluetooth, to exchange the CSI information between users. In all cases, the exchanged information should be reduced as much as possible without affecting the efficiency of the scheduling process.

- After receiving the CSI of all other users, each user implements the same low complexity scheduling algorithm, Algorithm 9.1 derived in Section 9.3. Hence, each user will know which subcarriers are allocated to it and which are allocated to other users. Finally, each user communicates with the CCD over its allocated subcarriers until the next scheduling interval starts.

Although greedy scheduling achieves a higher system rate than PF, PF scheduling is more justified when collaborative distributed resource allocation is performed. In fact, edge users do not have any incentive to cooperate when greedy scheduling is applied, since they will spend resources in sending feedback information and then they will be unfavored by the scheduling process. However, with PF scheduling, they are motivated to cooperate since they will be achieving their Pareto optimal rate, as determined by the Nash bargaining solution. The importance of this result stems from the fact that users do not have to do actual "bargaining", that is, they do not have to iteratively take actions (transmitting on certain subcarriers with certain power) and then exchange information with the other users to inform them of the action taken until the equilibrium solution (NBS) is reached. In this case, users have to be aware of the CSI, transmit power, subcarrier allocation, of all other users on all subcarriers in each iteration of every scheduling interval. However, with the equivalence of PF and the NBS, all what the users need to achieve the Pareto optimal NBS is the implementation of a low complexity scheduling algorithm achieving proportional fairness. Such an algorithm is presented in Section 9.3. It does not rely on iterations, does not require the exchange of subcarrier and power allocation information, and can be easily implemented by mobile users since it has a linear complexity. The only information that needs to be exchanged is a reduced number of bits representing the quantized CSI on each subcarrier.

9.2.2 CSI Quantization Scheme

It is essential to have a reduced feedback rate between users. Hence, an efficient CSI quantization scheme must be used, since exchanging full CSI will lead to prohibitive feedback rates. Given N_b as the number of feedback bits per subcarrier ($N_b \geq 1$), and

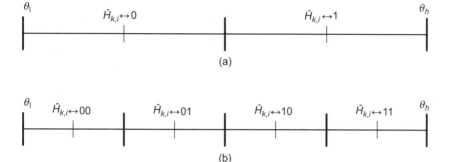

Figure 9.2 CSI quantization method. (a) 1-bit CSI; (b) 2-bit CSI. The symbol "↔" corresponds to the mapping of the quantization bits to the appropriate value $\hat{H}_{k,i}$ in $\mathcal{H}_q$. This figure is adapted from E. Yaacoub and Z. Dawy, "Achieving the Nash Bargaining Solution in OFDMA Uplink Using Distributed Scheduling with Limited Feedback", *International Journal of Electronics and Communication AEU (Elsevier)*, 65(4), 320–330, 2011.

θ_l and θ_h as the lower and upper allowed CSI values in dB, respectively, the step size (dB) can be determined as follows:

$$\theta_s = (\theta_h - \theta_l)/2^{N_b} \tag{9.1}$$

Users exchange the index of each quantized CSI value $\hat{H}_{k,i}$ selected from the set $\mathcal{H}_q$ defined by

$$\mathcal{H}_q = \left[\theta_l + \frac{\theta_s}{2}, \theta_l + \frac{3\theta_s}{2}, \cdots, \theta_h - \frac{\theta_s}{2} \right] \tag{9.2}$$

such that

$$\hat{H}_{k,i} = \arg \min_{\hat{H} \in \mathcal{H}_q} (|H_{k,i} - \hat{H}|) \tag{9.3}$$

where $|\cdot|$ denotes the absolute value operation. An example is shown in Fig. 9.2. The uniform quantization is performed on the CSI values in dB. This is justified by the fact that users are usually assumed to be uniformly distributed over the CCD coverage area, and thus their distance to the CCD is uniformly distributed. Furthermore, The fading distribution has a mean equal to the pathloss, which is generally proportional to $d^{-\upsilon}$, that is, the mean is proportional to a power of the distance. Hence, its expression in dB is proportional to the distance (due to the logarithm operation). Consequently, the value of the pathloss in dB is uniformly distributed in the range $[\theta_l \ \theta_h]$.

9.2.3 Price of Anarchy

A unified metric for quantifying the performance of the distributed scheduling schemes compared to the best achievable result (centralized scheduling) is desirable. Hence, to compare the distributed schemes to centralized schemes, the price of anarchy (PA), a frequently used metric in the literature, is applied. It is usually defined as the ratio of the cost of the suboptimal case to the cost of the optimal solution, for example, [146], or as the ratio of the maximum achievable utility to the utility achieved by the suboptimal solution [147]. The definition of Ref. [147] is best suited to the investigated approach. For PF, the utility to be used in PA is the logarithm of the rate, since it is also used in the scheduling algorithm. However, when the schemes are very close to optimal, the differences are not well expressed due to the slow increase of the logarithm operation. Hence, the product of the rates is considered as an additional metric for the PA, since the product is considerably more sensitive to the variations between the schemes. Hence, the two metrics are

$$
P_{A1} = \frac{\prod_{k} \left(R_k^{(c)}(P_k, \mathcal{I}_{\text{sub},k}) \right)}{\prod_{k} \left(R_k^{(d)}(P_k, \mathcal{I}_{\text{sub},k}) - O_k(\mathcal{I}_{\text{sub}}) \right)}
\tag{9.4}
$$

$$
P_{A2} = \frac{\sum_{k} \ln \left(R_k^{(c)}(P_k, \mathcal{I}_{\text{sub},k}) \right)}{\sum_{k} \ln \left(R_k^{(d)}(P_k, \mathcal{I}_{\text{sub},k}) - O_k(\mathcal{I}_{\text{sub}}) \right)}
\tag{9.5}
$$

where $R^{(c)}$ is the rate in the centralized case, $R^{(d)}$ is the rate in the distributed case, and O_k is the overhead in bps incurred by user k to transmit its CSI over the subcarriers in $\mathcal{I}_{\text{sub}}$. It is given by $O_k = N_b N_{\text{sub}}/T$, with T the duration of one TTI. As the ratios in (9.4) and (9.5) become close to 1, the performance becomes closer to optimal. The two metrics will have different numerical values, but both will lead to similar conclusions regarding the performance of the different schemes.

9.3 DISTRIBUTED SCHEDULING ALGORITHM

In this section, a linear complexity algorithm that constitutes an extension of Algorithm 7.2 to the case of distributed scheduling with quantized CSI, is presented. Algorithm 7.2 is a centralized scheduling algorithm for the full CSI case. In this section, Algorithm 9.1 is presented as a low complexity scheduling algorithm that can be implemented in centralized and distributed scenarios with full or quantized CSI. Algorithm 9.1 is an extension of Algorithm 7.2 to the quantized CSI case. The algorithm can be used with any utility function while having the property of ensuring proportional fairness when the utility used is the logarithm of the rate. Algorithm 9.1 consists of allocating RB i to user k in a way to maximize the difference

$$
\Lambda_{k,i} = U_k \left(R_k(P_k, \mathcal{I}_{\text{RB},k} \cup \{i\}) | \hat{\mathbf{H}} \right) - U_k \left(R_k(P_k, \mathcal{I}_{\text{RB},k}) | \hat{\mathbf{H}} \right)
\tag{9.6}
$$

The utility $U_k(R_k|\hat{\mathbf{H}})$ of user k in a quantized CSI scenario depends on the rate R_k given the matrix $\hat{\mathbf{H}}$ of the quantized CSI values $\hat{H}_{k,j}$ corresponding to user k over subcarrier j.

Algorithm 9.1

Linear Complexity Scheduling Algorithm for Scheduling with Quantized CSI

Subcarrier Allocation:
Consider the set of users $\mathcal{I}_{\text{users}} = \{1, 2, ..., K\}$ **sorted from 1 to** K **according to IDs allocated sequentially by the CCD when each user joins the network.**

> **for** k **such that** $1 \leq k \leq K$
> $\quad \mathcal{I}_{\text{RB},k} \leftarrow \emptyset$
> **end for**
> **for** i **such that** $1 \leq i \leq N_{\text{RB}}$
> $\quad$ **for all** k **such that** $1 \leq k \leq K$
> $\quad\quad$ **Compute** $U_k(P_k, \mathcal{I}_{\text{RB},k} \cup \{i\}|\hat{\mathbf{H}})$
> $\quad\quad$ **if** $\mathcal{I}_{\text{RB},k} \neq \emptyset$
> $\quad\quad\quad$ **Compute** $U_k(P_k, \mathcal{I}_{\text{RB},k}|\hat{\mathbf{H}})$
> $\quad\quad$ **else**
> $\quad\quad\quad U_k(P_k, \mathcal{I}_{\text{RB},k}|\hat{\mathbf{H}}) \leftarrow 0$
> $\quad\quad$ **end if**
> $\quad\quad \Lambda_{k,i} \leftarrow U_k(P_k, \mathcal{I}_{\text{RB},k} \cup \{i\}|\hat{\mathbf{H}}) - U_k(P_k, \mathcal{I}_{\text{RB},k}|\hat{\mathbf{H}})$
> $\quad$ **end for**
> $\quad\quad k^* \leftarrow \arg\max_k \Lambda_{k,i}$. **If more than one user satisfy** $k^* = \arg\max_k \Lambda_{k,i}$, **allocate** i **to the user having the lowest ID in** $\mathcal{I}_{\text{users}}$. **This convention is applied instead of arbitrary allocation, in order to ensure a consistent application of the algorithm by all users when the algorithm is implemented in a distributed way. In the centralized case, the BS can break ties arbitrarily.**
> $\quad$ **if** $\Lambda_{k^*,i} > 0$
> $\quad\quad \mathcal{I}_{\text{RB},k^*} \leftarrow \mathcal{I}_{\text{RB},k^*} \cup \{i\}$
> $\quad$ **end if**
> **end for**
> **Power Allocation:**
> **Apply Algorithm 6.2 for power allocation**

Algorithm 9.1 has a linear complexity. In fact, the algorithm allocates each RB after performing a linear search on the users in order to find the user that maximizes the marginal utility. Consequently, the total complexity of the algorithm is $\mathcal{O}(N_{\text{RB}}K)$, that is, the algorithm has linear complexity in the number of users and in the number of RBs, and thus could be implemented in real-time. In addition, Algorithm 9.1 results in a unique allocation. In fact, due to the quantized CSI, several users might have equal values for the marginal utility (9.6). However, Algorithm 9.1 solves this problem by sorting the users according to their IDs and allocating the RB to the user with the lowest ID.

9.3.1   Rate Calculations with Quantized CSI

For the rate calculations with quantized CSI in Algorithm 9.1, the following expression is used

$$R(P_k, \mathcal{I}_{\text{sub},k} | \hat{\mathbf{H}}) = \sum_{i=1}^{N_{\text{sub}}} \alpha_{k,i} \frac{B}{N} \cdot \log_2 \left(1 + \beta \hat{\gamma}_{k,i} \right) \tag{9.7}$$

$\hat{\gamma}_{k,i}$ is the estimated SNR of user k over subcarrier i in the case of quantized CSI $\hat{H}_{k,i}$. It is obtained by replacing, in (7.13) $\gamma_{k,i}$ by $\hat{\gamma}_{k,i}$ and $H_{k,i}$ by $\hat{H}_{k,i}$.

In the cooperative distributed scenario described in Section 9.2, users use the expression in (9.7) while implementing the scheduling algorithm since they only know the quantized CSI of other users. However, after subcarrier allocation, the rate achieved by each user is a function of its actual (not quantized) CSI, given by (7.12). Conversely, in the centralized case, the BS uses the expression in (7.12) while implementing the scheduling algorithm, or, equivalently, (9.7) with $\hat{H}_{k,i} = H_{k,i}$ since the BS is assumed to know the exact CSI values.

9.4   RESULTS AND DISCUSSION

This section presents the simulation results obtained by applying Algorithm 9.1 with different feedback bits per subcarrier.

9.4.1   Simulation Model

The simulation model consists of a single CCD with users located within a radius of 100 m from the CCD. This radius is relatively limited in order to allow the signals transmitted by users on the cell edge to be within the reference sensitivity level of users at the diametrically opposed edge (the reference sensitivity level is discussed in Ref. [148]). Users are located at fixed distances from the CCD, such that they are uniformly distributed in the interval [0 100] meters. The total bandwidth of 5 MHz is subdivided into 16 RBs as in Ref. [41]. In order not to have excessive complexity in the mobile terminals, one subcarrier per RB is assumed. This allows reducing the feedback overhead without compromising accuracy. In fact, fading can be assumed approximately constant within the bandwidth of a single RB since the scheme presented in this chapter is applicable in scenarios with limited or no mobility. A target BER of 10^{-6} is considered. The maximum user transmit power is considered to be 125 mW. All users are assumed to transmit at the maximum power, and the power is subdivided equally among all subcarriers allocated to the user. The channel gain over subcarrier i corresponding to user k is given by (7.15) with the same parameter values as in Section 7.5.1. The rate is averaged over 100 TTIs, with the duration of a TTI being 1 ms. Then the simulation is repeated over 100 iterations.

The threshold θ_l is selected such that $P_k 10^{\theta_l/10}$ corresponds to the sensitivity level in Ref. [148], and θ_h is selected such that $P_k 10^{\theta_h/10}$ corresponds to the maximum receiver power in Ref. [148], as shown in Table 9.1.

TABLE 9.1   Threshold Values

Receiver sensitivity level	-103 dBm
Max. receiver power	-25 dBm
$P_{k,\max}$	21 dBm
θ_l	-124 dBm
θ_h	-46 dBm

9.4.2   Greedy Scheduling Results

This section presents the results of greedy scheduling with quantized feedback ($U = R$). Fig. 9.3 shows the rate results as a function of the distance when four users are considered in the system, and Fig. 9.4 shows the corresponding average subcarrier allocation for each of the four users. Similarly, Fig. 9.5 shows the rate results as a function of the distance when eight users are considered in the system, and Fig. 9.6 shows the corresponding average subcarrier allocation for each of the eight users.

Figs 9.3–9.6 show that as the number of feedback bits increases, the results become closer to the full CSI (centralized) case. However, results close to the centralized case can be achieved with three or even two bits feedback per subcarrier. The case of one bit feedback is relatively far from optimal. Despite this fact, it can be seen that it has the side effect of favoring edge users at the expense of the users nearest to the CCD. This side effect can be considered desirable in practice, since it ensures more fairness to edge users while keeping a relative advantage for the nearest users.

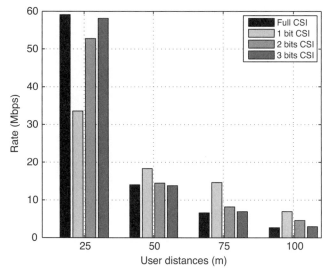

Figure 9.3   Rate achieved by each user as a function of the distance from the CCD—Greedy scheduling: four users. This figure is adapted from E. Yaacoub and Z. Dawy, "Achieving the Nash Bargaining Solution in OFDMA Uplink Using Distributed Scheduling with Limited Feedback", *International Journal of Electronics and Communication AEU (Elsevier)*, 65(4), 320–330, 2011.

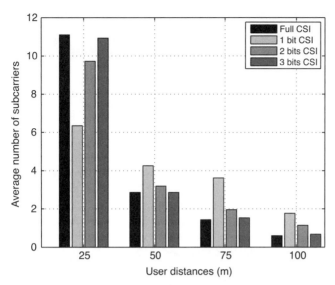

Figure 9.4 Average subcarrier allocation versus distance—Greedy scheduling: four users. This figure is adapted from E. Yaacoub and Z. Dawy, "Achieving the Nash Bargaining Solution in OFDMA Uplink Using Distributed Scheduling with Limited Feedback", *International Journal of Electronics and Communication AEU (Elsevier)*, 65(4), 320–330, 2011.

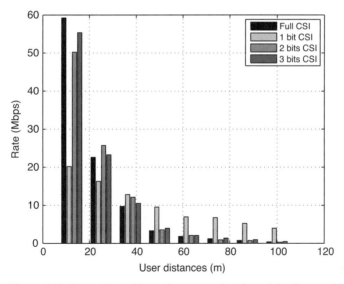

Figure 9.5 Rate achieved by each user as a function of the distance from the CCD—Greedy scheduling: eight users. This figure is adapted from E. Yaacoub and Z. Dawy, "Achieving the Nash Bargaining Solution in OFDMA Uplink Using Distributed Scheduling with Limited Feedback", *International Journal of Electronics and Communication AEU (Elsevier)*, 65(4), 320–330, 2011.

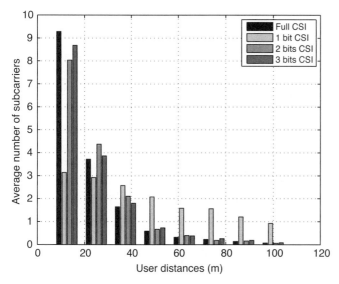

Figure 9.6   Average subcarrier allocation versus distance—Greedy scheduling: eight users. This figure is adapted from E. Yaacoub and Z. Dawy, "Achieving the Nash Bargaining Solution in OFDMA Uplink Using Distributed Scheduling with Limited Feedback", *International Journal of Electronics and Communication AEU (Elsevier)*, 65(4), 320–330, 2011.

9.4.3   PF Scheduling Results

Algorithm 9.1 is implemented by the users. Hence, only the PFF scheme is used in the distributed scheduling scenario, since it is not practical to assume that each user is capable of keeping track of the previously achieved rate of all other users in order to implement PFTF scheduling. In fact, the users have an estimation of the rate of other users according to (9.7) in order to implement the scheduling algorithm, but only the user knows its actual achieved rate according to (7.12). Hence, the only exchanged information is a fixed reduced number of feedback bits per subcarrier.

Figs 9.7 and 9.8 show the rate and subcarrier allocation results for four users, whereas Figs 9.9 and 9.10 show the rate and subcarrier allocation results for eight users. The results of the centralized PFF and PFTF1 schemes (which achieved the best performance in the centralized case studied in Chapter 7) are shown for comparison. These figures show that near-optimal results can be achieved with only one bit feedback. As the number of feedback bits increases, the results become closer to optimal. However, the very slight increase in rate indicates that one bit feedback can be used efficiently. The near-optimal results do not justify the investigation of more than three feedback bits per subcarrier. The achieved results with such a low number of feedback bits are due to the efficient mapping of bits to quantized values as described by the approach of Section 9.2.2.

Tables 9.2 and 9.3 show the price of anarchy results for the four and eight users case, respectively. The metrics defined in (9.4) and (9.5) are used. Since only the PFF scheme was implemented with quantized CSI, a fair comparison should be

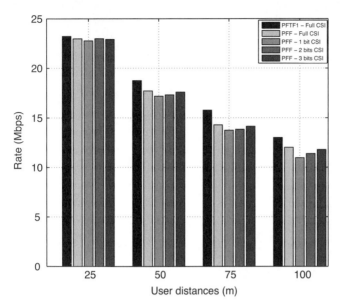

Figure 9.7 Rate results as a function of the distance from the CCD: distributed PFF scheduling with four users and 16 subcarriers. This figure is adapted from E. Yaacoub and Z. Dawy, "Achieving the Nash Bargaining Solution in OFDMA Uplink Using Distributed Scheduling with Limited Feedback", *International Journal of Electronics and Communication AEU (Elsevier)*, 65(4), 320–330, 2011.

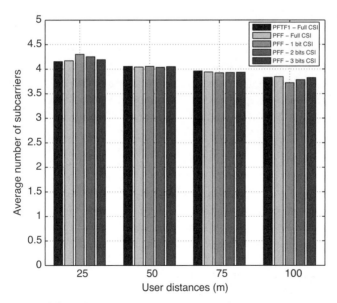

Figure 9.8 Subcarrier allocation results as a function of the distance from the CCD: distributed PFF scheduling with four users and 16 subcarriers. This figure is adapted from E. Yaacoub and Z. Dawy, "Achieving the Nash Bargaining Solution in OFDMA Uplink Using Distributed Scheduling with Limited Feedback", *International Journal of Electronics and Communication AEU (Elsevier)*, 65(4), 320–330, 2011.

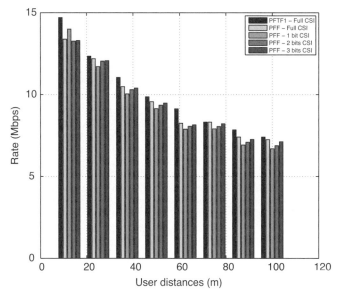

Figure 9.9   Rate results as a function of the distance from the CCD: distributed PFF scheduling with eight users and 16 subcarriers. This figure is adapted from E. Yaacoub and Z. Dawy, "Achieving the Nash Bargaining Solution in OFDMA Uplink Using Distributed Scheduling with Limited Feedback", *International Journal of Electronics and Communication AEU (Elsevier)*, 65(4), 320–330, 2011.

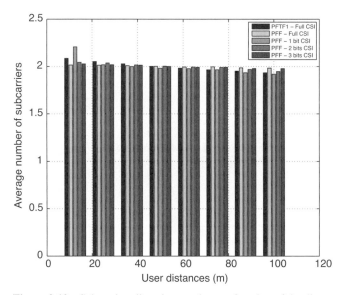

Figure 9.10   Subcarrier allocation results as a function of the distance from the CCD: distributed PFF scheduling with eight users and 16 subcarriers. This figure is adapted from E. Yaacoub and Z. Dawy, "Achieving the Nash Bargaining Solution in OFDMA Uplink Using Distributed Scheduling with Limited Feedback", *International Journal of Electronics and Communication AEU (Elsevier)*, 65(4), 320–330, 2011.

TABLE 9.2 Price of Anarchy Results for Distributed Scheduling: Four Users Case

Method	$\sum\limits_{k=1}^{K} R_k$ (Mbps)	$\prod\limits_{k=1}^{K} R_k$ $\times 10^{-4}$	$\sum\limits_{k=1}^{K} \ln(R_k)$	O_k (kbps)	$P_{A1}^{(PFF)}$	$P_{A2}^{(PFF)}$	$P_{A1}^{(PFTF)}$	$P_{A2}^{(PFTF)}$
PFTF-Full	70.73	8.92	66.66	–	–	–	–	–
PFF-Full	66.98	6.98	66.42	–	–	–	–	–
PFF-1 bit	64.66	5.89	66.25	16	1.1895	1.0026	1.5202	1.0063
PFF-2 bits	65.52	6.27	66.31	32	1.1230	1.0017	1.4352	1.0054
PFF-3 bits	66.43	6.72	66.38	48	1.0518	1.0008	1.3442	1.0045

This table is reprinted, with permission from Elsevier, from E. Yaacoub and Z. Dawy, "Achieving the Nash Bargaining Solution in OFDMA Uplink Using Distributed Scheduling with Limited Feedback", *International Journal of Electronics and Communication AEU (Elsevier)*, 65(4), 320–330, 2011 ©2011 Elsevier.

with the centralized PFF scheme with full CSI. Furthermore, since the logarithmic utility is used in the algorithm, (9.5) is more representative of the actual performance. However, a comparison with the PFTF1 scheme is also presented for reference since it represents the long-term optimal NBS. In addition, since the variations of the ln function are small when the performance is close between the different scenarios, (9.4) is also used in the comparisons. Although different values of the price of anarchy are yielded by the two metrics, they both lead to the same conclusions: the quantized CSI schemes perform closer to PFF than PFTF1, and the performance becomes closer to the full CSI case as the number of feedback bits increases.

Although the results of Figs 9.7–9.10 correspond to a comparison of the different cases with the centralized schemes at each user location and hence are more representative than P_{A1} and P_{A2}, the price of anarchy results represent a unified metric for quantifying the performance of each approach.

TABLE 9.3 Price of Anarchy Results for Distributed Scheduling: Eight Users Case

Method	$\sum\limits_{k=1}^{K} R_k$ (Mbps)	$\prod\limits_{k=1}^{K} R_k$ $\times 10^{-7}$	$\sum\limits_{k=1}^{K} \ln(R_k)$	O_k (kbps)	$P_{A1}^{(PFF)}$	$P_{A2}^{(PFF)}$	$P_{A1}^{(PFTF)}$	$P_{A2}^{(PFTF)}$
PFTF-Full	80.67	8.74	128.81	–	–	–	–	–
PFF-Full	76.87	6.04	128.44	–	–	–	–	–
PFF-1 bit	74.30	4.35	128.11	16	1.4095	1.0027	2.0404	1.0056
PFF-2 bits	75.06	4.88	128.23	32	1.2742	1.0019	1.8446	1.0048
PFF-3 bits	76.02	5.49	128.34	48	1.1483	1.0011	1.6623	1.0040

This table is reprinted, with permission from Elsevier, from E. Yaacoub and Z. Dawy, "Achieving the Nash Bargaining Solution in OFDMA Uplink Using Distributed Scheduling with Limited Feedback", *International Journal of Electronics and Communication AEU (Elsevier)*, 65(4), 320–330, 2011 ©2011 Elsevier.

9.5 SUMMARY

Distributed uplink scheduling in OFDMA systems was considered. Cooperation between mobile users was implemented using limited feedback of channel state information. Results close to optimal were achieved with a limited number of feedback bits. Furthermore, only one bit feedback per subcarrier was sufficient to achieve near-optimal results with proportional fair scheduling, which represents the Pareto optimal solution in distributed scheduling. In the next chapter, distributed scheduling without user cooperation will be investigated.

CHAPTER 10

DISTRIBUTED SCHEDULING WITHOUT USER COOPERATION

In Chapter 9, distributed scheduling with user cooperation was investigated. This chapter treats the noncooperative scenario, and is organized as follows. Section 10.1 presents the background information and lists the main ideas presented in the chapter. A distributed scheduling scheme is described in Section 10.2. It is based on channel sensing and probabilistic transmission. A comparison to other existing schemes is presented in Section 10.3. The impact of measurement inaccuracies is analyzed in Section 10.4. In Section 10.5, Simulation results are presented and discussed. The optimization of the transmission probabilities is investigated in Section 10.6, and some practical aspects of the presented scheme are discussed in Section 10.7. Finally, the chapter summary is presented in Section 10.8.

10.1 BACKGROUND

The benefits of distributed resource allocation are being widely investigated, since mobile devices have more autonomy in making transmission decisions in distributed schemes. In addition to CR, ad hoc, and sensor networks [33, 53–63], which are distributed in nature, distributed resource allocation has also been implemented in infrastructure based networks where users are connected to a central BS. In fact, several standards for 3G CDMA cellular networks, for example, 1xEV-DO [64, 65], have introduced mechanisms that give users greater independence in making transmission decisions best matched to their applications, for example, deciding when to transmit and at what rate. The cost of this flexibility is potentially higher intracell interference, and a corresponding degradation in performance [149, 150]. To maintain some control over the interference in 1xEV-DO, base stations set a Reverse-link Activity Bit (RAB) on when the load increases, and clear it in the opposite case [151]. The RAB is sometimes used in conjunction with a token bucket, where tokens are in units of transmission power, and the base station controls the token generation rate and the token bucket depth [149]. The uplink channel MAC design of cdma2000 1xEV-DO Rev A system is discussed in Ref. [152], where an algorithm is presented to update the token generation rate and the token bucket depth based on the computation of

Resource Allocation in Uplink OFDMA Wireless Systems: Optimal Solutions and Practical Implementations,
Elias E. Yaacoub and Zaher Dawy.
© 2012 by the Institute of Electrical and Electronics Engineers, Inc. Published 2012 by John Wiley & Sons, Inc.

the RAB. The uplink rate control algorithm for cdma2000 is presented in Ref. [153], where transition from one rate to another is performed by mobile devices in a probabilistic manner based on the value of the RAB. A rate control algorithm inspired from the cdma2000 1xEV-DO rate control algorithm is proposed for High Speed Uplink Packet Access (HSUPA) in Ref. [154], where it is shown that the proposed algorithm outperforms the conventional scheduling algorithm used in UMTS Release 6. Since the transmission rates are varied in a probabilistic manner, a study of the optimal transition probabilities (optimal in the sense of maximizing the sum-rate) is presented in Ref. [149] in a basic scenario where a simple on–off scheduler is studied, that is, users either transmit at a fixed rate R with a probability p, or do not transmit with a probability $(1 - p)$. Optimization of the transmission probabilities is also treated in Refs [155–157].

In this chapter, a distributed scheduling approach based on probabilistic transmission in conjunction with spectrum sensing is presented for OFDMA-based systems. In the investigated scheme, users are grouped into priority levels depending on their channel state information on each subcarrier. Users with higher priorities are given the privilege of transmitting before their counterparts. However, in order to allow fair access to wireless resources, probabilistic transmission is adopted. The transmission probabilities are optimized in order to achieve a desired performance in terms of rate and fairness. Compared to other access schemes in the literature, the presented scheme avoids many of their drawbacks, while possessing several advantages, including robustness to collisions and to channel measurement errors.

With broadband wireless access systems gaining more importance, users are expecting ubiquitous and seamless access to a variety of bandwidth demanding services. Mobile devices capable of supporting multiple standards are becoming more common in the market. Current research is not only ongoing on enhancing scheduling techniques within a given network, but also on optimizing the resource allocation in heterogenous networks. This involves selecting the best network to serve a user, among several networks with completely different access technologies such as GSM/EDGE, UMTS/HSPA, WiMAX, and WLAN [50–52]. The presented scheme allows users, subscribed in an OFDMA-based cellular network, to use the same radio interface to access an OFDMA based LAN, thus using the same accessing method for both long-range and short-range communications. This is in contrast with current terminals using several radio access technologies to simultaneously support, for example, both the CDMA-based UMTS and the various 802.11 standards, thus using different accessing methods for long-range and short-range communications. Furthermore, the studied scheme allows users to act as primary users in their home network or as secondary users in a cognitive radio network outside their home network, while using the same functionality in each network. Thus, the presented approach represents a step toward convergence over the air interface with OFDMA as the method of wireless access.

The main topics discussed in this chapter are summarized as follows:

1. Presenting a distributed scheduling approach for OFDMA uplink without user cooperation. The scheme is based on channel sensing and probabilistic transmission.

2. Investigating the performance of the presented approach with various transmission probabilities, in addition to studying the possibility of optimizing these probabilities.

3. Presenting a discussion of the benefits of the studied scheme compared to other schemes in the literature, and showing that it can overcome many of their limitations.

4. Discussing the practical implementation aspects of the presented approach, for example, its application in DBS systems and CR networks.

10.2 NONCOOPERATIVE DISTRIBUTED SCHEDULING SCHEME

10.2.1 System Model

A distributed resource allocation scheme based on channel sensing and probabilistic transmission is presented. By probabilistic transmission, it is meant that when a subcarrier is allocated to a user, the user either transmits with a certain probability p or refrains from transmission with probability $1 - p$. Channel sensing requires that users be able to sense the channel and "hear" the transmissions of other users, for example, in a CR scenario as in Refs [58, 61]. Therefore, the system studied consists of a single CCD covering an area of interest. The coverage area is assumed small enough so that the users can effectively sense the channel and listen to the transmissions of other users. The CCD can have the same practical implementations discussed in Section 9.2.1. The performance of the presented scheme is investigated within the range of a single CCD.

10.2.2 Distributed Scheduling Scheme

The noncooperative distributed scheduling scheme is based on slots of N_{TTI} TTIs each of duration T. The first TTI is reserved for subcarrier allocation and the remaining $(N_{TTI} - 1)$ TTIs are dedicated to user transmissions. The steps of the distributed scheme are shown in Fig. 10.1. The scheme can be described as follows:

- The CCD transmits a pilot signal over each of the subcarriers to be allocated for a period of duration $T/4$. The users can be synchronized with the CCD as follows: when a user joins the network, it waits until the next cycle of N_{TTI} TTIs begins, that is, it waits until it senses the pilot transmission from the CCD. It should be noted that the pilot signals transmitted by the CCD correspond to subcarriers used for uplink transmission and the approach can be used in FDD or TDD modes. In the case of FDD, the pilot signals correspond to uplink subcarriers, although they are transmitted by the CCD, and are hence orthogonal to the subcarriers used by the CCD for downlink transmission. In the case of TDD, the pilot transmission stage is considered as part of the uplink transmission time, and thus no downlink data transmission is allowed on the subcarriers during this stage.

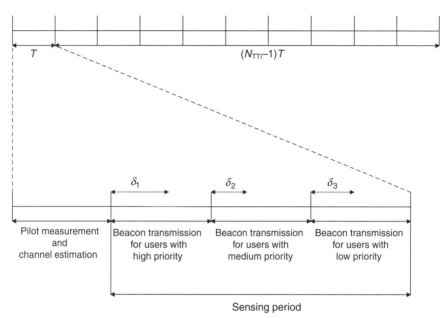

Figure 10.1 Steps of the distributed scheduling scheme. This figure is reprinted, with permission from IEEE, from E. Yaacoub and Z. Dawy, "Distributed Probabilistic Scheduling in OFDMA Uplink using Subcarrier Sensing", IEEE Wireless Communications and Networking Conference (WCNC 2009), Budapest, Hungary, April 2009 ©2009 IEEE.

- Based on the received signal level, users perform channel estimation on each subcarrier and determine their CSI. Based on their CSI, each user associates itself with a priority group based on quantized levels of estimated SNRs, since a higher CSI on a certain subcarrier leads to a higher achievable SNR on that subcarrier. In this chapter, three priority levels are considered. Letting $\Delta_{k,i}$ denote the pilot level received by user k on subcarrier i, and θ_l and θ_h denote lower and upper thresholds, respectively, then

    1. If $\Delta_{k,i} \geq \theta_h$, user k estimates that it is capable of achieving high SNR on subcarrier i and associates itself to priority group 1.
    2. If $\theta_l \leq \Delta_{k,i} < \theta_h$, user k estimates that it is capable of achieving moderate SNR on subcarrier i and associates itself to priority group 2.
    3. If $\Delta_{k,i} < \theta_l$, user k estimates that it is capable of achieving low SNR on subcarrier i and associates itself to priority group 3.

    The selection of the threshold values is discussed in Section 10.5.

- After the pilot measurement phase ends, a scheduling phase of duration $T/4$ is dedicated to users with the highest priority (priority 1). For each subcarrier i, each user k with priority 1 waits for a random delay $\delta_{1;k,i} < T/4$ while sensing the channel at the same time. If a signal on subcarrier i is sensed before $\delta_{1;k,i}$ has elapsed, then user k knows that another user k' belonging to priority group 1 has reserved the subcarrier since it had $\delta_{1;k',i} < \delta_{1;k,i}$. If, on the other hand,

$\delta_{1;k,i}$ elapsed and user k did not sense any transmission on subcarrier i, it then transmits a beacon signal on i with probability p_1, or does not transmit with probability $1 - p_1$.

- After the first scheduling phase of duration $T/4$ ends, a second scheduling phase of similar duration is dedicated to users belonging to priority level 2. For each subcarrier i that was not allocated during the previous phase, each user k with priority 2 waits for a random delay $\delta_{2;k,i} < T/4$ while sensing the channel at the same time. If a signal on subcarrier i is sensed before $\delta_{2;k,i}$ has elapsed, then user k knows that the subcarrier was reserved by another user. If, on the other hand, $\delta_{2;k,i}$ elapsed and user k did not sense any transmission on subcarrier i, it then transmits a beacon signal on i with probability p_2, or does not transmit with probability $1 - p_2$.

- After the second scheduling phase of duration $T/4$ ends, a third scheduling phase of similar duration is dedicated to users belonging to priority level 3. For each subcarrier i that was not allocated during the previous two phases, each user k with priority 3 waits for a random delay $\delta_{3;k,i} < T/4$ while sensing the channel at the same time. If a signal on subcarrier i is sensed before $\delta_{3;k,i}$ has elapsed, then user k knows that the subcarrier was reserved by another user. If, on the other hand, $\delta_{3;k,i}$ elapsed and user k did not sense any transmission on subcarrier i, it then transmits a beacon signal on i with probability p_3, or does not transmit with probability $1 - p_3$.

- After all three scheduling phases have ended (first TTI), the transmission of beacon signals stops and actual transmission of user data begins for a duration of $(N_{TTI} - 1)$ TTIs. Beacon transmission is only meant to reserve the subcarrier for a given user until the scheduling phase ends. The user then transmits its own data during the transmission phase on the subcarriers allocated to it during the resource allocation phase.

It should be noted that the number of scheduling phases was set to 3 based on a trade-off between the scheduling fairness and the duration of the scheduling interval: a large number of scheduling phases would make their overall duration too long and leave less time available for transmission. A shorter number would lead to unfairness for users having good channel conditions since users having a broader range of channel variations will be in the same category.

10.3   COMPARISON TO EXISTING SCHEMES

In this section, the similarities and differences between the scheme presented in this chapter and several schemes in the literature are described. In the presented scheme, transmission of pilot signals occurs in a periodic sequence similarly to beacon transmission in predefined intervals in IEEE 802.15.4 [158], where it is succeeded by a contention period followed by a contention free period in which guaranteed time slots are allocated to certain users. However, in the presented scheme, the scheduling phase can be considered as a contention period, but for resource allocation, not transmission. The transmission phase is always contention free. The division of users into priority levels provides a more simple approach than in elimination yield nonpreemptive multiple access (EY-NMPA) used in HiperLAN 1 [159], where several phases of sensing and accessing the medium for contention resolution are interleaved before one user is allowed to access the medium for data transmission [160]. In carrier sense multiple access (CSMA), users sense the channel and transmit when it is idle. With p-persistent CSMA, users transmit with a probability p [161]. Sensing and probabilistic transmission are essential to the presented scheme, but they occur at the resource allocation phase. Once the subcarriers are allocated, they are dedicated to the user for the whole transmission phase. This process has some similarity to the demand assigned multiple access (DAMA) used to enhance the classical Aloha [162]. DAMA consists of a reservation phase where users contend for dedicated time slots, and a transmission phase where each user transmits on the time slot it reserved, if any. In the presented scheme, collisions might occur only during the resource allocation phase not during the transmission phase, similarly to DAMA. However, users reserve subcarriers for the whole transmission period, not time slots. In addition, subdivision into priority groups and probabilistic transmission contribute in reducing the collision probability. Furthermore, the random delays used in each of the three scheduling intervals also contribute to reducing collisions, contrarily to CSMA with collision avoidance (CSMA/CA) where backoff takes place after a collision has occurred. The presented scheme avoids the overhead introduced by request to send/clear to send (RTS/CTS) packets in addition to other delays (e.g., SIFS, DIFS, etc.) introduced by multiple access with collision avoidance (MACA) used in conjunction with CSMA/CA in WLANs. In inhibit sense multiple access (ISMA), used in cellular digital packet data (CDPD) in the AMPS system, the BS uses a busy tone to indicate that the medium is busy [163]. In the presented scheme, users transmit the beacon in the resource allocation phase to ensure the reservation of a particular subcarrier. The scheme could be modified, without any change in the analysis and the obtained results, so that this transmission is done by the CCD after the user signals the reservation of the subcarrier to the CCD (this issue is discussed in Section 10.7.3).

The most comparable standards to the presented scheme are 802.11a and HyperLAN 2, which both use OFDM. HyperLAN 2 has most of the functionalities of 802.11a in addition to some more advanced features [160]. In fact, the subdivision of a single TTI into four phases has a similar counterpart in HyperLAN 2, although with completely different purposes, since in HyperLAN 2 a 2 ms frame is divided into broadcast, downlink, uplink, and random access phases, each with a variable duration [164–167]. Although no dynamic allocation of subcarriers to users is available in HyperLAN 2 (contrarily to the presented scheme), the AP senses the medium and dynamically selects the appropriate frequency within its coverage area (802.11h ensures similar functionality to 802.11a). In addition, mobility and handover between APs and interoperability with 3G are possible.

10.4 ANALYSIS OF MEASUREMENT INACCURACIES

In this section, the measurement inaccuracies are analyzed and the probability that a user misclassifies its priority order due to errors in measuring its CSI is determined.

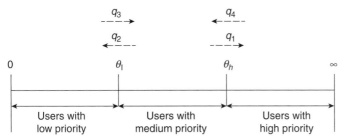

Figure 10.2   Probability of measurement errors.

Without loss of generality, it is assumed in this section that the CSI measurements are normalized with respect to the noise power and the transmitted pilot power. This allows simplifying the derivations since estimating $\Delta_{k,i}$ when the pilot power and noise power are known is equivalent to estimating $H_{k,i}$, the channel gain over subcarrier i when allocated to user k. The quantities investigated are enumerated below:

1. Probability $q_{k,i,1}$ that $H_{k,i}$ is in the interval $[\theta_l, \theta_h]$ and the user estimates it is in $]\theta_h, \infty[$. The average of $q_{k,i,1}$ over $[\theta_l, \theta_h]$ will be denoted by q_1.

2. Probability $q_{k,i,2}$ that $H_{k,i}$ is in the interval $[\theta_l, \theta_h]$ and the user estimates it is in $[0, \theta_l[$. The average of $q_{k,i,2}$ over $[\theta_l, \theta_h]$ will be denoted by q_2.

3. Probability $q_{k,i,3}$ that $H_{k,i}$ is in the interval $[0, \theta_l[$ and the user estimates it is in $[\theta_l, \infty[$. The average of $q_{k,i,3}$ over $[0, \theta_l[$ will be denoted by q_3.

4. Probability $q_{k,i,4}$ that $H_{k,i}$ is in the interval $]\theta_h, \infty[$ and the user estimates it is in $[0, \theta_h]$. The average of $q_{k,i,4}$ over $]\theta_h, \infty[$ will be denoted by q_4.

The different cases are shown in Fig. 10.2. The error $e_{k,i}$ in estimating $H_{k,i}$ is modeled by a Gaussian random variable with zero-mean and variance σ_e^2. Assuming Gaussian estimation error was also adopted in Refs [113–115]. This model is justified by the fact that the users should be able to make measurements that are more likely to be close to the actual value of the CSI and less likely to be too far from the actual value. It is also justified by the fact that many well-designed estimators are asymptotically Gaussian [115]. Consequently, the following expressions can be easily derived:

$$q_{k,i,1} = \text{Prob}\left(H_{k,i} + e_{k,i} > \theta_h | H_{k,i} \in [\theta_l, \theta_h]\right)$$
$$= Q\left(\frac{\theta_h - H_{k,i}}{\sigma_e}\right) \tag{10.1}$$

$$q_{k,i,2} = \text{Prob}\left(H_{k,i} + e_{k,i} < \theta_l | H_{k,i} \in [\theta_l, \theta_h]\right)$$
$$= Q\left(\frac{H_{k,i} - \theta_l}{\sigma_e}\right) \tag{10.2}$$

$$q_{k,i,3} = \text{Prob}\left(H_{k,i} + e_{k,i} > \theta_l | H_{k,i} \in [0, \theta_l]\right)$$
$$= Q\left(\frac{\theta_l - H_{k,i}}{\sigma_e}\right) \tag{10.3}$$

$$q_{k,i,4} = \text{Prob}\left(H_{k,i} + e_{k,i} < \theta_h | H_{k,i} \in [\theta_h, \infty[\right)$$

$$= Q\left(\frac{H_{k,i} - \theta_h}{\sigma_e}\right) \tag{10.4}$$

where $Q(x) = \int_x^\infty \frac{1}{\sqrt{2\pi}} e^{-y^2/2} dy$. An alternate form of the Q-function implied by the work of Ref. [168] and defined for $x \geq 0$ is given by [169]

$$Q(x) = \frac{1}{\pi} \int_0^{\pi/2} e^{-\frac{x^2}{2\sin^2(\phi)}} d\phi \tag{10.5}$$

The advantage in the form given by (10.5) is that the variable is not in the integrand, and the limits of the integration are finite. This makes the form in (10.5) suitable for numerical integration. This alternate form will be used in the following computations. The channel gain $H_{k,i}$ in Rayleigh fading is modeled as an exponential random variable with mean $\overline{H}_{k,i}$ and probability density function $P_H(H_{k,i}) = \frac{1}{\overline{H}_{k,i}} e^{-H_{k,i}/\overline{H}_{k,i}}$ [169]. The indices k, i will be dropped to simplify the notations when no confusion can occur. Hence, the following is obtained:

$$q_1 = \int_{\theta_l}^{\theta_h} \text{Prob}(H + e > \theta_h) P_H(H) dH$$

$$= \int_{\theta_l}^{\theta_h} Q\left(\frac{\theta_h - H}{\sigma_e}\right) \cdot \frac{1}{\overline{H}} e^{-H/\overline{H}} dH \tag{10.6}$$

$$= \frac{1}{\pi \overline{H}} \int_{\theta_l}^{\theta_h} \int_0^{\pi/2} e^{\frac{-\overline{H}H^2 + 2H(\theta_h \overline{H} - \sigma_e^2 \sin^2(\phi)) - \theta_h^2 \overline{H}}{2\overline{H}\sigma_e^2 \sin^2(\phi)}} d\phi dH$$

$$q_2 = \int_{\theta_l}^{\theta_h} \text{Prob}(H + e < \theta_l) P_H(H) dH$$

$$= \int_{\theta_l}^{\theta_h} Q\left(\frac{H - \theta_l}{\sigma_e}\right) \cdot \frac{1}{\overline{H}} e^{-H/\overline{H}} dH \tag{10.7}$$

$$= \frac{1}{\pi \overline{H}} \int_{\theta_l}^{\theta_h} \int_0^{\pi/2} e^{\frac{-\overline{H}H^2 + 2H(\theta_l \overline{H} - \sigma_e^2 \sin^2(\phi)) - \theta_l^2 \overline{H}}{2\overline{H}\sigma_e^2 \sin^2(\phi)}} d\phi dH$$

$$q_3 = \int_0^{\theta_l} \text{Prob}(H + e > \theta_l) P_H(H) dH$$

$$= \int_0^{\theta_l} Q\left(\frac{\theta_l - H}{\sigma_e}\right) \cdot \frac{1}{\overline{H}} e^{-H/\overline{H}} dH \tag{10.8}$$

$$= \frac{1}{\pi \overline{H}} \int_0^{\theta_l} \int_0^{\pi/2} e^{\frac{-\overline{H}H^2 + 2H(\theta_l \overline{H} - \sigma_e^2 \sin^2(\phi)) - \theta_l^2 \overline{H}}{2\overline{H}\sigma_e^2 \sin^2(\phi)}} d\phi dH$$

$$q_4 = \int_{\theta_h}^{\infty} \text{Prob}(H + e < \theta_h) P_H(H) dH$$

$$= \int_{\theta_h}^{\infty} Q\left(\frac{H - \theta_h}{\sigma_e}\right) \cdot \frac{1}{\overline{H}} e^{-H/\overline{H}} dH \qquad (10.9)$$

$$= \frac{1}{\pi \overline{H}} \int_{\theta_h}^{\infty} \int_{0}^{\pi/2} e^{\frac{-\overline{H}H^2 + 2H(\theta_h \overline{H} - \sigma_e^2 \sin^2(\phi)) - \theta_h^2 \overline{H}}{2\overline{H}\sigma_e^2 \sin^2(\phi)}} d\phi dH$$

Fig. 10.3 shows the numerical evaluation of (10.6) to (10.9). In this figure, the error standard deviation σ_e is taken as a proportion of the mean channel gain $\overline{H}$ (or equivalently the mean SNR since the values are normalized). The range of values considered for $\overline{H}$ encompasses the interval from the user sensitivity level till the user maximum input level as defined in Ref. [148]. Although the upper limit of the outer integral in (10.9) is infinity, it can still be easily evaluated numerically by integrating up to $10\overline{H}$ since $\int_{0}^{10\overline{H}} \frac{1}{\overline{H}} e^{-H/\overline{H}} dH \approx 1 = \int_{0}^{\infty} \frac{1}{\overline{H}} e^{-H/\overline{H}} dH$. The threshold values used are $\theta_l = -92$ dB and $\theta_h = -75$ dB. The selection of the thresholds was based on the range of values obtained with the propagation model considered in addition to the reference sensitivity level of the users [148].

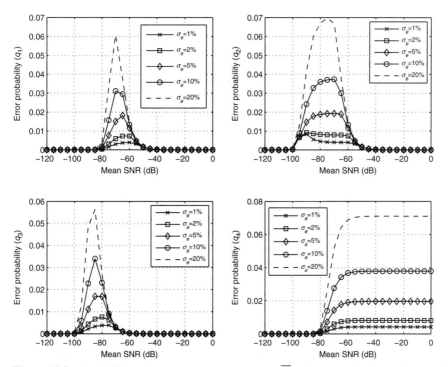

Figure 10.3 Probabilities of measurement errors versus $\overline{H}$ with $\theta_l = -92$ dB and $\theta_h = -75$ dB. Upper left: q_1, upper right: q_2, lower left: q_3, and lower right: q_4.

At very low mean SNRs, all the error probabilities go to zero, since the estimation error must be extremely large to make the estimated SNR greater than θ_l in the case of q_3, and the probability of H being greater than θ_l when $\overline{H}$ is far less than θ_l is relatively small in the case of q_1, q_2, and q_4, due to the nature of the exponential distribution. For very high mean SNRs, the probability of H being less than θ_h when $\overline{H}$ is considerably greater than θ_h is relatively small in the case of q_1, q_2, and q_3. However, in the case of q_4, the error is large enough (since it is considered proportional to $\overline{H}$) to shift the measured value below θ_h with non-negligible probability. Furthermore, after a certain value of the mean SNR, the error probability reaches a stable "floor" since the outer integration goes to infinity in (10.9) and the exponential distribution decays fast when the sample values are too far from the mean.

However, in all cases, the error probability does not exceed 7%, when σ_e reaches a value as high as 20% of $\overline{H}$. In addition, the effect of the measurement errors, when they occur, is not critical neither on the system operation nor on the ability of users to transmit successfully. In fact, the subcarrier might even not be allocated to the user where the measurement error occurred, due to the random delays and probabilistic transmission, since all users of the same priority group with respect to a certain subcarrier have equal probability of being allocated that subcarrier. In this case no loss in system rate (or fairness) will occur. In addition, if a user misses being allocated a subcarrier, it can increase the power allocated to its other subcarriers that leads to a partial compensation of the loss of the particular subcarrier. Consequently, due to the low probability of measurement errors, in addition to their limited effects when they occur, perfect CSI estimation will be assumed in the sequel.

10.5   RESULTS AND DISCUSSION

In this section, the performance of the distributed scheduling scheme presented in Section 10.2 is evaluated via Monte Carlo simulations. Results related to the rate and fairness for various sets of transmission probabilities and different numbers of users within the coverage range of the CCD are presented. Rate calculations are performed according to (7.12). The SNR over a single subcarrier is given by (7.13). The power is assumed to be subdivided equally among all the subcarriers allocated to a user. The transmission at the maximum power is done to maximize the user rate, since the subcarriers are orthogonal and the coverage area of a single CCD is considered in this chapter (i.e., interference is not an issue).

10.5.1   Simulation Model

The simulation model consists of a single CCD with users located within a radius of 100 m from the CCD. The range selection is comparable to the range of existing standards, for example, WLAN or Bluetooth that has a 100 m range for a 100 mW transmit power [170]. Scheduling is performed in slots of 10 TTI intervals, with the first interval used for resource allocation and the remaining nine TTIs used for

TABLE 10.1   Transmission Probabilities

	Probabilities		
	p_1	p_2	p_3
Case 1	1	1	1
Case 2	0.5	0.5	0.5
Case 3	0.75	0.5	0.25
Case 4	0.75	0.5	1

transmission. The duration of a TTI is assumed to be 1 ms. The simulation is repeated over 10,000 iterations. The total bandwidth considered is $B = 5$ MHz, subdivided into 16 subcarriers as in Ref. [41]. A target BER of 10^{-6} and a maximum user transmit power of 125 mW are considered. The channel gain over subcarrier i corresponding to user k is given by (7.15) with the same parameter values as in Section 7.5.1.

In the simulations, users are located at fixed distances from the CCD, such that they are equidistant in the interval [0 100] meters (e.g., four users are positioned at 25, 50, 75, and 100 m). The thresholds considered are $\theta_l = -92$ dB and $\theta_h = -75$ dB relative to the pilot power transmitted by the CCD. The selection of the thresholds was based on the range of values obtained with the propagation model considered in addition to the reference sensitivity level of the users [148]. Different sets of transmission probabilities (defined in Section 10.2.2) are investigated. The various cases are listed in Table 10.1.

Case 1 represents a greedy approach where in every scheduling phase, a user transmits on every available subcarrier when the corresponding timer $\delta_{k,i}$ expires. Case 2 represents a fair approach where the probability of transmission for each user on each subcarrier is equal to the probability of leaving the subcarrier available for other users. In Case 3, users having good CSI on a subcarrier are given a higher probability to access this subcarrier whereas users with lower CSI have a lower chance of accessing it. Case 4 applies the concept of Case 3 but tries to avoid leaving unallocated subcarriers by setting $p_3 = 1$.

10.5.2   Simulation Results

Fig. 10.4 shows the rate results as a function of the distance when four users are considered in the system, and Fig. 10.5 shows the corresponding average subcarrier allocation for each of the four users. Similarly, Fig. 10.6 shows the rate results as a function of the distance when eight users are considered in the system, and Fig. 10.7 shows the corresponding average subcarrier allocation for each of the eight users. Figs 10.8 and 10.9 show the rate and subcarrier allocation results, respectively, in the case of 16 users. The results obtained by using the noncooperative distributed scheduling scheme are compared to centralized scheduling using greedy and PFF utilities in Algorithm 7.2.

The figures show that when the distributed scheme is used with the transmission probabilities of Case 1, its behavior is similar to greedy scheduling: most resources are allocated to the users close to the CCD whereas users at large distances suffer

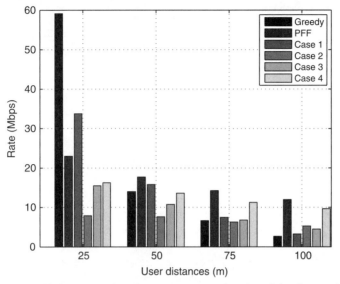

Figure 10.4 Rate achieved by each user as a function of the distance from the base station: four users case. This figure is reprinted, with permission from IEEE, from E. Yaacoub and Z. Dawy, "Distributed Probabilistic Scheduling in OFDMA Uplink using Subcarrier Sensing", IEEE Wireless Communications and Networking Conference (WCNC 2009), Budapest, Hungary, April 2009 ©2009 IEEE.

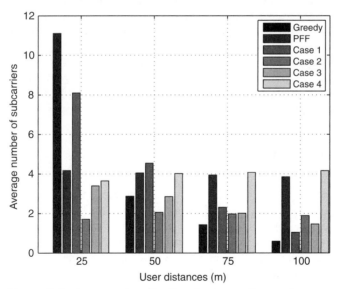

Figure 10.5 Average subcarrier allocation versus distance: four users case. This figure is reprinted, with permission from IEEE, from E. Yaacoub and Z. Dawy, "Distributed Probabilistic Scheduling in OFDMA Uplink using Subcarrier Sensing", IEEE Wireless Communications and Networking Conference (WCNC 2009), Budapest, Hungary, April 2009 ©2009 IEEE.

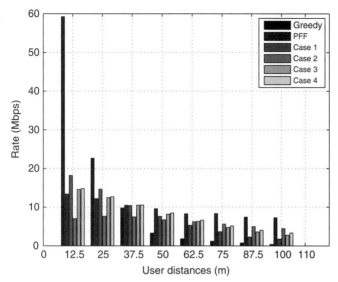

Figure 10.6 Rate achieved by each user as a function of the distance from the base station: eight users case. This figure is reprinted, with permission from IEEE, from E. Yaacoub and Z. Dawy, "Distributed Probabilistic Scheduling in OFDMA Uplink using Subcarrier Sensing", IEEE Wireless Communications and Networking Conference (WCNC 2009), Budapest, Hungary, April 2009 ©2009 IEEE.

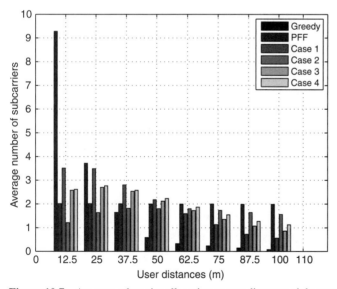

Figure 10.7 Average subcarrier allocation versus distance: eight users case. This figure is reprinted, with permission from IEEE, from E. Yaacoub and Z. Dawy, "Distributed Probabilistic Scheduling in OFDMA Uplink using Subcarrier Sensing", IEEE Wireless Communications and Networking Conference (WCNC 2009), Budapest, Hungary, April 2009 ©2009 IEEE.

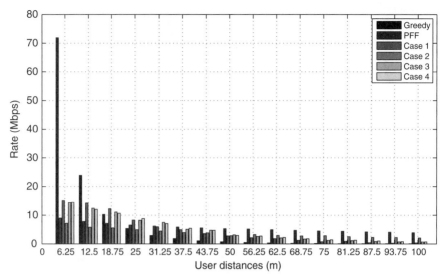

Figure 10.8 Rate achieved by each user as a function of the distance from the base station: 16 users case. This figure is reprinted, with permission from IEEE, from E. Yaacoub and Z. Dawy, "Distributed Probabilistic Scheduling in OFDMA Uplink using Subcarrier Sensing", IEEE Wireless Communications and Networking Conference (WCNC 2009), Budapest, Hungary, April 2009 ©2009 IEEE.

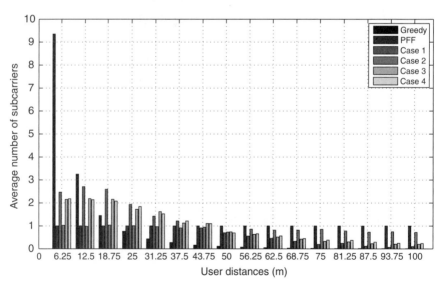

Figure 10.9 Average subcarrier allocation versus distance: 16 users case. This figure is reprinted, with permission from IEEE, from E. Yaacoub and Z. Dawy, "Distributed Probabilistic Scheduling in OFDMA Uplink using Subcarrier Sensing", IEEE Wireless Communications and Networking Conference (WCNC 2009), Budapest, Hungary, April 2009 ©2009 IEEE.

from starvation. This is due to the fact that all the probabilities in Case 1 are equal to one. This allows users with good channels on certain subcarriers to reserve those subcarriers with probability one, thus depriving users with worse channels from being allocated these subcarriers. Case 2 leads to a high degree of fairness since all users are allocated on average almost equal numbers of subcarriers, and their achieved average rates are also almost identical. This can be explained by the fact that all the transmission probabilities are set to 0.5. In this situation, users with better channels have equal chances of transmitting or of leaving the subcarriers to be allocated for other users with lower channel qualities. Contrarily to Case 1, the scenario of Case 2 ensures maximum fairness at the expense of achieving low network rate. Case 3 has similar behavior to PF (by "similar" it is meant that they have the same trend in their performance, not necessarily the same values): it allocates more resources than Case 1 to users located further away from the CCD, at the expense of a decrease in rate for nearby users. Case 4 has very close performance to Case 3 for users having good and medium channel levels. However, it differs from it in that it tries to avoid having unallocated subcarriers by setting $p_3 = 1$, thus providing more resources for edge users.

It should be noted that, as the number of users increases, the performance of Cases 3 and 4 becomes closer to that of PF scheduling, in terms of rate and fairness, as can be seen in Figs 10.4–10.9. In addition, except for the few users that are the nearest to the CCD (2 users out of 8 in Figs 10.6 and 10.7, and 2 users out of 16 in Figs 10.8 and 10.9), all four cases of the distributed scheme outperform centralized greedy scheduling in terms of rate and fairness for all the remaining users.

The presented approach scales well with the number of users. However, it is normally not possible to increase the number of users beyond the investigated limits, due to the limited coverage area requirement for applying the distributed approach.

10.6 OPTIMIZATION OF TRANSMISSION PROBABILITIES

After investigating distributed probabilistic scheduling with different sets of transmission probabilities, it is clear that the performance is affected by the selected transmission probabilities. Consequently, optimizing the transmission probabilities in order to achieve desired performance goals in terms of rate and/or fairness will be the subject of this section. Maximizing the cell rate can be achieved by simply maximizing the sum of the rates $\sum_k R(P_k, \mathcal{I}_{\text{sub},k})$ over the set of probabilities. However, in order to achieve proportional fairness among the users, the sum of the logarithm of the rates $\sum_k \ln \left(R(P_k, \mathcal{I}_{\text{sub},k}) \right)$ should be maximized over the set of probabilities. Hence, the optimization over the sum of two utility functions is considered: the rate, compared to Case 1 (since Case 1 represents a greedy approach), and the logarithm of the rate, compared to Case 4 (since Case 4 represents the closest performance to proportional fairness among the studied cases).

10.6.1 Optimization Methods

The probabilities are optimized in a two-step process using the simulated annealing (SA) technique in the first step and genetic algorithms (GA) in the second. SA employs a random search by applying an arbitrary perturbation on the parameters of an objective function u. As the algorithm moves from one state to another, not only does it keep all the perturbations that increase u, but also it accepts the perturbations that decrease it with a probability $e^{-\Delta u/T_s}$, where Δu is the decrease in u due to the perturbation, and T_s is a control parameter. By accepting some changes that decrease u, the SA algorithm avoids being trapped in local maxima. In the studied case, the objective function (or utility) to maximize is either $\sum_k R(P_k, \mathcal{I}_{\text{sub},k})$, denoted by "greedy optimization" in the results, or $\sum_k \ln \left(R(P_k, \mathcal{I}_{\text{sub},k}) \right)$, denoted by "fair optimization" in the results. In both cases, with the presented scheme, the function can only be evaluated via simulation as in Section 10.5, due to the use of random delays and probabilistic transmission. The SA algorithm was run for 10 times to optimize each of the utilities. Each run of SA led to a different set of probabilities, although the value of the utility was almost the same in all runs (less than 1% fluctuation). This case can occur when a function has more than one global maximum, or when it has several local maxima of values slightly less than the value of the global maximum. To reach a single optimal set of probabilities, GA was applied with the initial population being the 10 outcomes of the SA algorithm. Each outcome consisting of a vector $[p_1, p_2, p_3]$ becomes a member of the population. New populations are generated by cloning, crossover, and mutation between the probability vectors, then the best probability vector is selected as the solution. Cloning is obtained by duplicating the same vector in the population, that is, by creating a new vector $[p_1', p_2', p_3']$ such that $[p_1' = p_1, p_2' = p_2, p_3' = p_3]$. Crossover between two vectors is obtained by taking certain elements from one vector, and the remaining elements from the other. for example, given two vectors $[p_1, p_2, p_3]$ and $[p_1', p_2', p_3']$, a vector $[p_1'', p_2'', p_3'']$ with crossover using two parameters from the first vector and one parameter from the second can be obtained such that $[p_1'' = p_1, p_2'' = p_2, p_3'' = p_3']$. Finally, mutation is obtained by inducing a small random perturbation on one of the vector elements. Since the elements are probabilities, it should be taken into account that the perturbed probability should still be in the interval [0, 1].

10.6.2 Optimization Results

The optimization was performed for four and eight users. The optimal probabilities are shown in Table 10.2. In the greedy optimization, it can be seen for the four users case that p_1 and p_2 are high in order to allow users with good channels to transmit. It can be noted that p_3 is relatively lower. This can be explained by the fact that for users with an extremely bad channel it is sometimes more beneficial not to transmit on that channel and subdivide the power on its other allocated channels. In the eight users case, p_1 and p_2 are relatively lower since the increase of the number of users increases the probability that an edge user has a relatively good channel on certain subcarriers. This mandates that users with better channels transmit with a lower probability in order to give more chances to the edge users. In the fair optimization, p_1 is less than in the greedy optimization for both four and eight users, in order to allow more available subcarriers to users suffering from lower CSI and hence achieve more fairness.

TABLE 10.2   Optimized Transmission Probabilities

Scenario	Number of users	p_1	p_2	p_3
Greedy optimization	4	0.9892	0.8265	0.6431
Fair optimization	4	0.7942	0.7853	0.9435
Greedy optimization	8	0.8428	0.6120	0.9137
Fair optimization	8	0.7052	0.6969	0.7337

It should be noted that, in all the cases of Table 10.2, although none of the probabilities is set to 1, the average number of allocated subcarriers is greater than 15.9, that is, the optimized probabilities lead to the allocation of all the subcarriers except in some limited extreme scenarios, as discussed above for cell edge users.

Table 10.3 shows the values of the utility functions in the investigated scenarios with and without optimization. The utilities shown are the sum-rate in Mbps, and the sum of the logarithms of the rate. It should be noted that, for a given number of users, the difference between the results of the last column of Table 10.3 is relatively small due to the logarithm operation. The results of the centralized scheduling schemes (greedy and PFF) are also shown for comparison. As expected, they outperform the distributed scheduling scheme. Comparing the distributed scheduling scenarios, it can be seen from Table 10.3 that the greedy optimization achieves the highest sum-rate

TABLE 10.3   Comparison of the Optimized and Nonoptimized Scenarios

Scenario	Number of users	$\sum_k R(P_k, \mathcal{I}_{\text{sub},k})$ (Mbps)	$\sum_k \ln\left(R(P_k, \mathcal{I}_{\text{sub},k})\right)$
Case 1	4	60.270	64.7278
Case 2	4	26.999	62.8484
Case 3	4	37.472	63.7846
Case 4	4	50.774	65.3487
Greedy optimization	4	60.629	64.7752
Fair optimization	4	55.562	65.5285
Centralized Greedy	4	82.4380	64.8512
Centralized PFF	4	66.9775	66.4152
Case 1	8	63.822	124.8261
Case 2	8	50.043	125.0622
Case 3	8	63.141	125.8734
Case 4	8	65.583	126.3670
Greedy optimization	8	67.987	125.8798
Fair optimization	8	64.654	126.5021
Centralized Greedy	8	99.0312	120.6858
Centralized PFF	8	76.8654	128.4404

results whereas the fair optimization achieves the best results in terms of the sum of logarithms of the rate. In the four users scenario, Case 1 achieves the closest results to the greedy optimization, whereas it achieves relatively close results in the eight users scenario (Case 4 is closer than Case 1 by less than 3%). On the other hand, Case 4 achieves the closest performance to the fair optimization in both the four and eight users scenarios.

Although the results of Section 10.5.2 represent a comparison of the different cases with the centralized schemes at each user location, a unified metric for quantifying the performance of each approach is desirable. Hence, to compare the optimized distributed schemes to centralized schemes, the price of anarchy, discussed in Section 9.2.3, is used. However, in the noncollaborative approach presented in this chapter, the overhead O_k is incorporated in the user rate itself, since the time spent in the scheduling phase reduces the time available for transmission and hence lowers the achievable rate. Furthermore, the PA is used to assess greedy scheduling in addition to PFF scheduling, since in the noncollaborative approach users are not exactly implementing greedy and PFF utilities as in the collaborative approach of Chapter 9. Instead, they perform distributed resource allocation using the transmit probabilities set up by the CCD and communicated to the users at the beginning of the scheduling process. Users are not aware if the probabilities used are dedicated to achieving user fairness or a greedy maximization of the cell sum-rate. Hence, in this chapter, the price of anarchy P_A is defined as

$$
P_A = \frac{\sum_k U_c(P_k, \mathcal{I}_{\text{sub},k})}{\sum_k U_d(P_k, \mathcal{I}_{\text{sub},k})}
\tag{10.10}
$$

where U_c is the utility in the centralized case and U_d is the utility in the distributed case. Comparing the greedy optimization to the centralized greedy scheduling case, the utility used is the rate. On the other hand, the utility is the logarithm of the rate while comparing the fair optimization to the centralized PFF case. Using the utility values from Table 10.3, the price of anarchy results are displayed in Table 10.4.

As P_A becomes close to 1, the performance becomes closer to optimal. Hence, from Table 10.4, it can be seen that as the number of users increases, the greedy optimization approach diverges from the centralized sum-rate solution, but the fair optimization approach remains close to the centralized PFF solution. It should be

TABLE 10.4   Price of Anarchy of the Optimized Scenarios

Scenario	Number of users	P_A
Greedy optimization	4	1.3597
Fair optimization	4	1.0135
Greedy optimization	8	1.4566
Fair optimization	8	1.0153

noted that, the increase in P_A in the greedy optimization approach is mainly due to the contribution of the nearest one or two users in the sum-rate of the centralized greedy scheduling scheme. However, the greedy optimization scheme of the presented approach performs similarly to, or better than, centralized greedy scheduling for the remaining users, as seen in Section 10.5.2 (Figs 10.4 and 10.6).

10.7 PRACTICAL CONSIDERATIONS

In this section, some practical considerations related to the presented distributed scheme and its possible extensions are discussed.

10.7.1 Collisions

Let Δt be the duration required by a user to determine that a transmission is occurring over a certain subcarrier, with $\Delta t \ll T/4$. A collision occurs if, for a given subcarrier, the timers (δ) of two or more users expire within an interval Δt. Since the duration of each timer is uniformly distributed over an interval of duration $T/4$, the probability that a timer expires in a window of duration Δt is $4\Delta t/T$. The probability that the timers of M different users over the same subcarrier expire within the same interval Δt is $(4\Delta t/T)^M$. Obviously, $(4\Delta t/T)^M \ll 1$, even for $M = 2$ since $\Delta t^2 \ll T^2$. Hence, the collision probability is very low. Besides, if two users detect a collision in the scheduling phase, they refrain from transmitting on the subcarrier where the collision occurred. Consequently, collisions may occur only in the scheduling phase, but not in the transmission phase. Detecting collisions at the scheduling phase allows avoiding simultaneous transmission and thus reception of corrupted data at the CCD. The effect is a slight reduction in system rate during the corresponding N_{TTI} slots, since the subcarrier is not used for transmission. However, this reduction is partly compensated since the power that should have been used for transmission on that subcarrier will be divided among the other successfully allocated subcarriers of the competing users, thus increasing their rate on those subcarriers. In addition, if a user can detect the identity of the other user competing on the same subcarrier (e.g., if direct sequence spread spectrum is used to encode the identity of a certain mobile on each beacon it transmits), the problem can be resolved by taking turns according to a certain predefined priority list. For example, a user higher in the list is allocated the subcarrier the first time a collision occurs. The other user is allocated the subcarrier in case a second collision occurs, and so on. The priority can simply be a number given to a user as it logs into the system.

10.7.2 Collaboration Between Mobile Users

In the presented scheme, there is no explicit cooperation between users. However, the classification into different priority levels, the random delays, and the probabilistic transmission could be considered as a form of "implicit" cooperation between users, since each user is performing resource allocation in a distributed way with minimal intervention from the CCD and by sensing the transmissions of other users. On the other hand, if direct or explicit cooperation between users were possible, a better performance would be achieved, as shown in Chapter 9. In fact, since users are able to sense the channel within the coverage area of interest, they could broadcast their CSI on the available subcarriers on some dedicated control channels other than the scheduled subcarriers. The CSI of a user will be sensed by the other users and the centralized scheduling algorithm can be run at each user instead of being run at the CCD. It should be noted that in Chapter 9, the CSI broadcasted by each user represents the channel state of the subcarriers between the CCD and the user, not between users. In other words, users are not performing any relaying. They exchange CSI to implement locally the resource allocation algorithm that would have been implemented at the CCD in the centralized scenario. Other collaborative techniques could include relaying of the data of a certain user by other users, or the formation of coalitions by groups of users so that they look as a single entity to the CCD.

10.7.3   Role of the Central Controlling Devices

Although the presented approach is based on distributed scheduling, the CCD still has a certain role in the resource allocation process. The pilot signals transmitted by the CCD at the beginning of each resource allocation phase provide a reference for new users joining the system to allow them to participate in the scheduling process. In addition, the presented scheme could be modified to give the CCD additional roles without hindering the distributed nature of the scheduling process. For example, when users reserve subcarriers, they can signal the reservation to the CCD who can then transmit the beacon signal instead of having the user transmit it by itself. This would constitute a possible solution to the collision scenario where the CCD decides which user is allocated a given subcarrier in case of collision.

10.7.4   Extension to a Single Cell Scenario

The studied model is in fact a building block in a more general scenario. In the presented model, users were assumed to be close enough from each other and from the CCD in order to successfully sense the wireless medium. The model can be incorporated into a more general framework where a central BS allocates resources to a certain number of CCDs spread geographically over the cell area, that is, the CCD could play the role of an RRH in a DBS scenario as described in Chapter 8. The distribution should be such that users associated with each CCD fall within a reasonable distance from the CCD so that they can successfully sense the air interface. The number of subcarriers to be allocated by the BS to a given CCD depends on the number of users associated with that CCD. A possible solution would be to allocate the subcarriers proportionally to the number of users associated with each CCD, as in Algorithm 8.1. After the subcarrier allocation phase, the performance analysis within the coverage area of each CCD is similar to the approach presented in this chapter.

10.7.5   Extension to a Multiple Cell Scenario

In a multiple cell scenario, the centralized BS in each cell allocates subcarriers to CCDs similarly to the single cell scenario. However, interference from neighboring cells should be taken into account during the resource allocation process. This can be achieved by having the CCDs in each cell periodically report the CSI information of their allocated subcarriers to the central BS. Having one or more subcarriers suffer from bad quality for a certain period of time is an indication that they are probably subjected to high interference. Consequently, they are reassigned to another CCD and other subcarrier(s) are assigned to the reporting CCD. This solution does not require cooperation between BSs. Another example of a decentralized interference management technique is presented in Refs [105, 171]. If, on the other hand, cooperation is possible, BSs can communicate to avoid allocating the same subcarriers to CCDs located at their common boundary. More details about distributed scheduling in a multicell scenario will be discussed in Chapter 12.

10.7.6   Cognitive Radio and 4G

In the studied scheme, the users are assumed to have similar capabilities to CR devices to be able to sense the wireless medium in order to determine the available subcarriers. In fact, the CCD could be part of a cellular network or a CR network. In a cellular network, the CCD could act as an RRH in a DBS system as discussed in Sections 10.7.4 and 10.7.5. In a CR network, the CCD could represent a device with higher capabilities than the user, and it could sense the medium for the presence of primary users and determine the available subcarriers and then announce these subcarriers to users within its coverage range by transmitting the appropriate pilot signals. In both cases, the users need the same functionalities and the same device can be used as a primary user in a cellular network or as a secondary user in a CR network. Hence, the presented approach could represent an important step for beyond-LTE and beyond-WiMAX 4G systems where the users could act as primary users in their home network and as secondary users in other networks without having to possess extra capabilities or use additional functionalities. Consequently, this chapter represents a step toward the convergence of wireless systems into an all CR network where users can move seamlessly between networks while using the same functionality in each network.

10.8   SUMMARY

In this chapter, distributed uplink scheduling in OFDMA systems without user cooperation was considered. Users were grouped into priority levels depending on their channel state information on each subcarrier. Users with higher priorities were given the privilege of transmitting before their counterparts. However, in order to allow fair access to wireless resources, probabilistic transmission was adopted. This allowed users with low priorities to have a reasonable chance of accessing the channel resources. Users determine if the channel is idle before transmission by sensing the medium over all the available subcarriers. It was shown through Monte Carlo simulations that the performance of the distributed scheduling mechanism leads to high rate results with reasonable fairness toward users that are far from the base station.

The transmission probabilities were optimized in order to achieve a desired performance in terms of rate and fairness. The presented scheme was proven to be robust to collisions and to channel measurement errors. It can be efficiently integrated in a multicell scenario with each cell containing a central BS connected to several CCDs spread throughout the cell. It also allows users to act as primary users or as secondary users in a cognitive radio network while using the same functionality in each network.

This chapter closes the part of the book dedicated to scheduling within a single cell. In Chapter 11, centralized resource allocation in a multicell scenario is studied, whereas Chapter 12 deals with distributed resource allocation in the multicell case.

CENTRALIZED MULTICELL SCHEDULING WITH INTERFERENCE MITIGATION

In Chapter 3, the problem of ergodic sum-rate maximization in the OFDMA uplink was formulated and solved. A single cell scenario was assumed, and the expectation was taken over the pdf of the CNR. In a multicell scenario, ergodic maximization should be taken over the pdf of the channel gain to interference plus noise ratio (CINR). However, with dynamic subcarrier and power allocation, the interference, and thus the CINR, vary dynamically with each scheduling decision taken. Therefore, instantaneous sum-rate maximization with centralized scheduling will be considered in this chapter. Section 11.1 presents the relevant background and lists the main topics of the chapter. The network sum-rate maximization problem is formulated in Section 11.2. The solution based on playing a pricing game between BSs is described in Section 11.3. The solution with a central control unit (CCU) enforcing the increase in network sum-rate is presented in Section 11.4. The suboptimal intercell interference mitigation/avoidance schemes with and without BS collaboration are discussed in Sections 11.5 and 11.6. The results are presented and analyzed in Section 11.7. Finally, the chapter is summarized in Section 11.8.

11.1 BACKGROUND

In the previous chapters, the focus was on a single cell scenario and hence intercell interference was not considered. To limit the effects of interference in multicell scenarios, several techniques for reusing the radio frequencies are investigated in the literature. Static reuse schemes are based on fractional frequency reuse (FFR) where a cell is divided into an inner area with the same frequencies reused in all cells and an outer area where a subset of the frequencies is reused [76]. More efficient schemes consist of dynamic frequency reuse where all the frequencies are allowed to be used in all cells and elaborate techniques are applied for interference mitigation or avoidance. In Ref. [79], downlink scheduling in OFDMA is used with each BS randomly turning on or off certain subcarriers to mitigate interference. In Ref. [80], down-

Resource Allocation in Uplink OFDMA Wireless Systems: Optimal Solutions and Practical Implementations,
Elias E. Yaacoub and Zaher Dawy.
© 2012 by the Institute of Electrical and Electronics Engineers, Inc. Published 2012 by John Wiley & Sons, Inc.

link transmit power allocation in a multicell wireless network under a sum-capacity maximization criterion and peak power constraints at each BS is investigated. It was shown that the optimal power control is binary (on–off) for two cells, and that binary power control yields a negligible capacity loss for more than two cells. However, scheduling is not used with power control in Ref. [80]. This has been explored in Ref. [81], where a distributed algorithm for power allocation and scheduling in multicell networks is proposed. Intercell coordination is applied to maximize the overall system capacity by deactivating cells that do not offer enough capacity to outweigh the interference caused to the network. A single frequency is considered in Ref. [81], and the application of the approach to OFDMA networks necessitates the use of subcarrier allocation to exploit the frequency diversity gain. In Ref. [82], resource allocation is considered in the downlink of multicell OFDMA systems without BS cooperation. A price that increases with the transmit power is used in order to reduce the interference. In Ref. [53], pricing is considered in ad hoc networks, where each user sets a price for other users to compensate for the interference they are causing. The prices are used as a sort of power control scheme to reduce transmission power. However, users are assumed to transmit on the same carriers and pricing is used for power control and not for scheduling. In Ref. [172], multicell uplink OFDMA scheduling is considered. Pricing is imposed by the network and each user performs power control in a distributed manner using the pricing information. Pricing is applied on the total transmit power of the user, not on the power transmitted on each individual subcarrier.

In this chapter, the network sum-rate maximization problem is formulated as a pricing game and an iterative solution is presented in the scenarios with and without central control. In addition, suboptimal scheduling schemes that do not rely on iterative techniques and applicable with and without BS cooperation are presented. The main areas of interest investigated in this chapter can be summarized as follows:

- Formulating the network sum-rate maximization problem as a pricing game played between the BSs and presenting an iterative solution based on utility maximization within each cell. The presented solution differs from Refs [53, 172] in that the transmit power is derived in closed form and the exclusivity of subcarrier allocation within each cell is explicitly enforced. Furthermore, scheduling is done by the BSs, not the users, since it is a stringent requirement to assume that users are aware of the CSI between all users and all the BSs in the network.

- Enhancing the solution obtained by the pricing game when a CCU is in charge of resource allocation in the network. In this case, each BS plays a double game: a pricing game with the other BSs in the network and a tit-for-tat game with the CCU. After each scheduling step taken by a BS, the CCU accepts the step if it leads to an increase in the total network rate, and rejects it if it leads to a selfish increase of the cell rate at the expense of the network rate.

- Presenting a suboptimal scheduling scheme based on pricing and collaboration between BSs. The scheme assumes that each BS knows only the CSI of the users within its coverage area and is not aware of their CSI with other BSs. In this scheme, BSs cooperate by exchanging interference information.

- Presenting a suboptimal scheduling scheme using pricing-based power control without collaboration between BSs. The presented scheme assumes that each BS knows only the CSI of the users within its coverage area and is not aware of their CSI with other BSs. In addition, each BS performs power control using a fixed price without exchanging information with other BSs.

- Presenting an interference avoidance scheme combined with uplink scheduling in OFDMA systems. The presented scheme does not require the exchange of pricing information, is not based on lengthy iterations, and does not require distributed power control. In addition, it does not involve BS cooperation. It is based on probabilistic on–off scheduling on each subcarrier within each cell depending on the received interference level.

- Comparing the results of the pricing games with and without CCU to the results of the suboptimal scheduling schemes with and without BS cooperation and showing that the suboptimal scheduling approach without BS cooperation achieves a comparable performance to the pricing game without CCU. This result has important practical implications, since it indicates that simple power control techniques can approximate the performance of complex methods based on the assumptions that all BSs communicate with each other for the purpose of scheduling and that all BSs are aware of the CSI of all users in the network.

11.2   PROBLEM FORMULATION

Letting N be the number of subcarriers, L the number of BSs, K_l the number of users in cell l, $P_{k_l,i,l}$ the power transmitted by user k_l over subcarrier i in cell l, P_{k_l} the total transmission power of user k_l, and $P_{k_l,\max}$ its maximum transmission power, the maximization of the weighted sum-rate in the network can be formulated as

$$\max_{\mathbf{P}} \sum_{l=1}^{L} \sum_{k_l=1}^{K_l} \sum_{i=1}^{N} \pi_{k_l} \log_2 \left(1 + \beta \frac{P_{k_l,i,l} H_{k_l,i,l}}{I_{i,l} + \sigma_{i,l}^2} \right) \tag{11.1}$$

subject to

$$0 \leq \sum_{i=1}^{N} P_{k_l,i,l} \leq P_{k_l,\max}; \forall k_l = 1, ..., K_l, \quad l = 1, ..., L \tag{11.2}$$

with

$$\mathbf{P} = \{\mathbf{P}_1, \mathbf{P}_2, ..., \mathbf{P}_l, ..., \mathbf{P}_L\} \tag{11.3}$$

where $\mathbf{P}_l$ is a $K_l \times N$ matrix of allocated powers $P_{k_l,i,l}$. The parameter β is the SNR gap given by (6.3). $H_{k_l,i,l}$ is the channel gain of user k_l over subcarrier i in cell l, and $\sigma_{i,l}^2$ is the noise power over subcarrier i in cell l.

The interference caused by user k_j in cell j over subcarrier i in cell l is given by

$$I_{k_j,i,l} = P_{k_j,i,j} H_{k_j,i,l} \tag{11.4}$$

The interference received over subcarrier i in cell l is given by

$$I_{i,l} = \sum_{j \neq l, j=1}^{L} \sum_{k_j=1}^{K_j} I_{k_j,i,l} \tag{11.5}$$

The weights π_{k_l} are used to give the rates of certain users more importance in the maximization of (11.1) thus providing a notion of fairness. They are chosen such that $\sum_{k_l=1}^{K_l} \pi_{k_l} = 1$. In an OFDMA-based wireless communication system, these weights are generally handed down from the MAC layer to the PHY layer scheduling routine on a per-frame (or longer) basis [26].

In order to reach the solution of the problem defined in (11.1), the Lagrangian, which entails a vector of Lagrange multipliers λ corresponding to the power constraint, is defined.

$$
\begin{aligned}
L(\mathbf{P}, \lambda) &= \sum_{l=1}^{L} \sum_{k_l=1}^{K_l} \sum_{i=1}^{N} \pi_{k_l} \log_2 \left(1 + \beta \frac{P_{k_l,i,l} H_{k_l,i,l}}{I_{i,l} + \sigma_{i,l}^2} \right) \\
&\quad - \sum_{l=1}^{L} \sum_{k_l=1}^{K_l} \lambda_{k_l} \left(\sum_{i=1}^{N} P_{k_l,i,l} - P_{k_l,\max} \right) \\
&= \sum_{l=1}^{L} \sum_{k_l=1}^{K_l} \sum_{i=1}^{N} \left\{ \pi_{k_l} \log_2 \left(1 + \beta \frac{P_{k_l,i,l} H_{k_l,i,l}}{I_{i,l} + \sigma_{i,l}^2} \right) - \lambda_{k_l} P_{k_l,i,l} \right\} \\
&\quad + \sum_{l=1}^{L} \sum_{k_l=1}^{K_l} \lambda_{k_l} P_{k_l,\max}
\end{aligned}
\tag{11.6}
$$

The formulation in (11.1) is not convex, since the interference term is a function of the power variables. However, it can be verified that the optimal solution will be regular, and hence the Karush–Kuhn–Tucker (KKT) conditions can be applied, even if the problem is not convex [35]. In fact, since in the uplink each user has its own power constraint regardless of the power constraints of other users, the gradients of the inequality constraints (11.2) are independent. Consequently, $\mathbf{P}$ is regular and the optimal solution $\mathbf{P}^*$ satisfies the KKT conditions [35]:

$$\frac{\partial L(\mathbf{P}, \lambda)}{\partial P_{k_l,i,l}^*} = 0 \tag{11.7}$$

Considering the normalized rate (with respect to the bandwidth) in bps/Hz of user k_l over subcarrier i in cell l,

$$r_{k_l,i,l} = \log_2 \left(1 + \beta \frac{P_{k_l,i,l} H_{k_l,i,l}}{I_{i,l} + \sigma_{i,l}^2} \right) \tag{11.8}$$

The KKT conditions can be written as

$$\frac{\partial L(,\mathbf{P},\lambda)}{\partial P_{k_l,i,l}} = \frac{\partial \sum_{l=1}^{L} \sum_{k_l=1}^{K_l} \sum_{i=1}^{N} \pi_{k_l} r_{k_l,i,l}}{\partial P_{k_l,i,l}} - \lambda_{k_l}$$

$$= \pi_{k_l} \frac{\partial r_{k_l,i,l}}{\partial P_{k_l,i,l}} + \sum_{m \neq l,m=1}^{L} \sum_{k_m=1}^{K_m} \pi_{k_m} \frac{\partial r_{k_m,i,m}}{\partial P_{k_l,i,l}} - \lambda_{k_l}$$

$$= \pi_{k_l} \frac{\partial r_{k_l,i,l}}{\partial P_{k_l,i,l}} + \sum_{m \neq l,m=1}^{L} \sum_{k_m=1}^{K_m} \pi_{k_m} \frac{\partial r_{k_m,i,m}}{\partial I_{i,m}} \frac{\partial I_{i,m}}{\partial P_{k_l,i,l}} - \lambda_{k_l}$$ (11.9)

$$= \pi_{k_l} \frac{\partial r_{k_l,i,l}}{\partial P_{k_l,i,l}} - \sum_{m \neq l,m=1}^{L} c_{i,m} \frac{\partial I_{i,m}}{\partial P_{k_l,i,l}} - \lambda_{k_l}$$

with

$$c_{i,m} = - \sum_{k_m=1}^{K_m} \pi_{k_m} \frac{\partial r_{k_m,i,m}}{\partial I_{i,m}}$$ (11.10)

Considering the first term of (11.9),

$$\frac{\partial r_{k_l,i,l}}{\partial P_{k_l,i,l}} = \frac{1}{\ln(2)} \frac{\frac{\beta H_{k_l,i,l}}{I_{i,l}+\sigma_{i,l}^2}}{1 + \frac{\beta P_{k_l,i,l} H_{k_l,i,l}}{I_{i,l}+\sigma_{i,l}^2}}$$

$$= \frac{1}{\ln(2)} \frac{\beta H_{k_l,i,l}}{I_{i,l} + \sigma_{i,l}^2 + \beta P_{k_l,i,l} H_{k_l,i,l}}$$ (11.11)

The second term of (11.9) can be developed as follows:

$$\frac{\partial r_{k_m,i,m}}{\partial I_{i,m}} = -\frac{1}{\ln(2)} \frac{\frac{\beta P_{k_m,i,m} H_{k_m,i,m}}{(I_{i,m}+\sigma_{i,m}^2)^2}}{1 + \frac{\beta P_{k_m,i,m} H_{k_m,i,m}}{I_{i,m}+\sigma_{i,m}^2}}$$

$$= -\frac{1}{\ln(2)} \frac{\beta P_{k_m,i,m} H_{k_m,i,m}}{(I_{i,m} + \sigma_{i,m}^2)^2 + (I_{i,m} + \sigma_{i,m}^2)\beta P_{k_m,i,m} H_{k_m,i,m}}$$ (11.12)

and

$$\frac{\partial I_{i,m}}{\partial P_{k_l,i,l}} = H_{k_l,i,m}$$ (11.13)

Using (11.11), (11.12), and (11.13) in (11.7), then

$$
P^*_{k_l,i,l} = \left[\frac{\pi_{k_l}}{\ln(2) \left(\displaystyle\sum_{m \neq l, m=1}^{L} c_{i,m} H_{k_l,i,m} + \lambda_{k_l} \right)} - \frac{I_{i,l} + \sigma^2_{i,l}}{\beta H_{k_l,i,l}} \right]^+
\tag{11.14}
$$

where $[y]^+ = \max(0, y)$.

The solution $P^*_{k_l,i,l}$ in (11.14) corresponds to the maximum sum-rate given the current interference conditions $I_{i,m}$ for $i = 1, ..., N$ and $m = 1, ..., L$. However, when $P^*_{k_l,i,l}$ is applied in cell l with subcarrier i allocated to user k_l, the interference $I_{i,m}$ in all $m \neq l$ will vary, thus leading to a variation in $c_{i,m}$, which necessitates a new $P^*_{k_l,i,l}$. Hence, the optimal transmit power in a given cell and the interference in all other cells are interwinded. Consequently, to solve this problem, the iterative approach described in Section 11.3 is used. It should be noted that in a single cell scenario, intercell interference is not an issue. This corresponds to setting $I_{i,l}$ and $c_{i,m}$ to zero in (11.14). In this case, the power allocation in (11.14) corresponds to the single cell water-filling solution obtained in Refs [37, 95].

11.3 ITERATIVE PRICING-BASED POWER CONTROL SOLUTION

At a given scheduling interval, the problem with KKT conditions (11.9) appears as a sum-rate maximization problem with pricing-based power control. In fact, $c_{i,m}$ plays the role of a price imposed by BS m on the user transmissions in other cells over subcarrier i. This price corresponds to the relative increase in the rate of cell m per unit decrease in the interference over subcarrier i. A similar interpretation was noted in Ref. [53] for ad hoc networks. However, the price $c_{i,m}$ in (11.14) depends on the power $P^*_{k_l,i,l}$. In order to solve this problem, the optimal solution for a single cell is presented given a fixed set of prices $\hat{c}_{i,m}$, then this solution is incorporated into an iterative pricing game between the L BSs, where each BS updates the value of $\hat{c}_{i,m}$ based on the actions taken by the other BSs. Consequently, in this section, the network sum-rate maximization problem is formulated as L single cell rate maximization problems subjected to pricing-based power control.

11.3.1 Single Cell Problem Formulation

The maximization of the weighted sum-rate with pricing in cell l can be formulated as

$$
\max_{\mathbf{A_l, P_l}} \sum_{k_l=1}^{K_l} \sum_{i=1}^{N} \left\{ \pi_{k_l} \alpha_{k_l,i,l} \log_2 \left(1 + \beta \frac{P_{k_l,i,l} H_{k_l,i,l}}{I_{i,l} + \sigma^2_{i,l}} \right) - \sum_{m \neq l, m=1}^{L} \hat{c}_{i,m} \alpha_{k_l,i,l} P_{k_l,i,l} H_{k_l,i,m} \right\}
\tag{11.15}
$$

subject to

$$0 \le \sum_{i=1}^{N} P_{k_l,i,l} \le P_{k_l,\max}; \quad \forall k_l = 1, ..., K_l, l = 1, ..., L \tag{11.16}$$

$$\sum_{k_l=1}^{K_l} \alpha_{k_l,i,l} \le 1; \quad \forall i = 1, ..., N, l = 1, ..., L \tag{11.17}$$

where $\mathbf{A_l}$ is a $K_l \times N$ matrix of channel allocation indices in cell l, $\alpha_{k_l,i,l}$. The sub-carrier allocation index $\alpha_{k_l,i,l} = 1$ if subcarrier i is allocated to user k_l in cell l, that is, $i \in \mathcal{I}_{\text{sub},k_l}$, with $\mathcal{I}_{\text{sub},k_l}$ the set of subcarriers allocated to user k_l in cell l. Otherwise, $\alpha_{k_l,i,l} = 0$.

The subcarrier allocation variables, $\alpha_{k_l,i,l}$, are used in addition to the power variables $P_{k_l,i,l}$, since the subcarriers are assumed to be allocated exclusively to each user in a single cell. Although the exclusivity of subcarrier allocation was implicitly assumed in the formulation of (11.1), since the interference was considered to come from users in other cells, the variables $\alpha_{k_l,i,l}$ are added to the formulation in (11.15) in order to make this exclusivity explicit. These variables facilitate the derivation of the optimal solution in Section 11.3.2 by ensuring a separability of the power and subcarrier allocation variables. Hence, (11.17) represents the constraint that each subcarrier can be allocated to a single user only in a given cell during one transmission time interval (TTI).

11.3.2 Single Cell Scheduling Solution

In order to facilitate the derivations, the condition on $\alpha_{k_l,i,l}$ is relaxed by assuming it can take any value in the interval $[0\ 1]$ rather than in the set $\{0, 1\}$. This corresponds to allowing time-sharing over the subcarriers.[3] However, it will be shown that the optimal solution will consist of exclusive subcarrier allocation for the whole duration of a TTI. Relaxing the condition on $\alpha_{k_l,i,l}$ and defining $\tilde{P}_{k_l,i,l} = \alpha_{k_l,i,l} P_{k_l,i,l}$, the maximization in (11.15) can be reformulated as follows:

$$\max_{\mathbf{A_l},\tilde{\mathbf{P}_l}} \sum_{k_l=1}^{K_l} \sum_{i=1}^{N} \left\{ \pi_{k_l} \alpha_{k_l,i,l} \log_2 \left(1 + \beta \frac{\tilde{P}_{k_l,i,l} H_{k_l,i,l}}{\alpha_{k_l,i,l}(I_{i,l} + \sigma_{i,l}^2)} \right) - \sum_{m \neq l,m=1}^{L} \hat{c}_{i,m} \tilde{P}_{k_l,i,l} H_{k_l,i,m} \right\} \tag{11.18}$$

subject to

$$0 \le \sum_{i=1}^{N} \alpha_{k_l,i,l} P_{k_l,i,l} = \sum_{i=1}^{N} \tilde{P}_{k_l,i,l} \le P_{k_l,\max}; \quad \forall k_l = 1, ..., K_l, l = 1, ..., L \tag{11.19}$$

---

[3] by "time-sharing", it is meant that several users can transmit on a given subcarrier during a given scheduling interval, with each user transmitting alone for a fraction of the interval. This corresponds to a sort of TDMA subdivision of the scheduling time unit and is not to be confused with overlapping transmission.

$$\sum_{k_l=1}^{K_l} \alpha_{k_l,i,l} \leq 1; \quad \forall i = 1, ..., N, l = 1, ..., L \tag{11.20}$$

The problem in (11.18) is a convex optimization problem, since the objective function is concave and the constraints are affine (a function of the form $a \log_2(1 + b/a)$ is known to be concave [34] and its sum with an affine function is concave).[4] The problem in (11.18) is equivalent to the original problem in (11.15) when the condition on the $\alpha_{k_l,i,l}$s is relaxed: for each user k_l and subcarrier i, finding the optimal pair $(\alpha^*_{k_l,i,l}, \tilde{P}^*_{k_l,i,l}) = (\alpha^*_{k_l,i,l}, \alpha^*_{k_l,i,l} P^*_{k_l,i,l})$ leads to the same solution as finding $(\alpha^*_{k_l,i,l}, P^*_{k_l,i,l})$. A similar decomposition to ensure problem convexity was adopted in Ref. [33]. Hence, finding the optimal dual solution corresponds to finding the optimal primal solution with zero duality gap. In order to reach the dual solution of the problem defined in (11.18), the Lagrangian is defined, which entails a vector of Lagrange multipliers λ corresponding to the power constraint.

$$L(\mathbf{A_l}, \mathbf{P_l}, \lambda) = \sum_{k_l=1}^{K_l} \sum_{i=1}^{N} \left\{ \pi_{k_l} \alpha_{k_l,i,l} \log_2 \left(1 + \beta \frac{P_{k_l,i,l} H_{k_l,i,l}}{I_{i,l} + \sigma^2_{i,l}} \right) - \sum_{m \neq l, m=1}^{L} \hat{c}_{i,m} \alpha_{k_l,i,l} P_{k_l,i,l} H_{k_l,i,m} \right\}$$
$$- \sum_{k_l=1}^{K_l} \lambda_{k_l} \left(\sum_{i=1}^{N} \alpha_{k_l,i,l} P_{k_l,i,l} - P_{k_l,\max} \right) \tag{11.21}$$

The Lagrangian dual function is given by

$$D(\lambda) = \max_{\mathbf{A_l}, \mathbf{P_l}} L(\mathbf{A_l}, \mathbf{P_l}, \lambda) \tag{11.22}$$

The optimization dual problem is given by

$$\min_{\lambda \geq 0} D(\lambda) \tag{11.23}$$

The Lagrangian dual can be rewritten as follows:

$$D(\lambda) = \max_{\mathbf{A_l}, \mathbf{P_l}} \sum_{k_l=1}^{K_l} \sum_{i=1}^{N} \alpha_{k_l,i,l}$$
$$\times \left\{ \pi_{k_l} \log_2 \left(1 + \beta \frac{P_{k_l,i,l} H_{k_l,i,l}}{I_{i,l} + \sigma^2_{i,l}} \right) - \sum_{m \neq l, m=1}^{L} \hat{c}_{i,m} P_{k_l,i,l} H_{k_l,i,m} - \lambda_{k_l} P_{k_l,i,l} \right\} + \sum_{k_l=1}^{K_l} \lambda_{k_l} P_{k_l,\max}$$
$$= \max_{\mathbf{A_l}} \sum_{k_l=1}^{K_l} \sum_{i=1}^{N} \max_{\mathbf{P_l}} \left\{ \alpha_{k_l,i,l} \Phi(P_{k_l,i,l}) \right\} + \sum_{k_l=1}^{K_l} \lambda_{k_l} P_{k_l,\max} \tag{11.24}$$

---

[4] In the limit, the following is verified: $\lim_{a \to 0} a \log_2(1 + b/a) = 0$.

with $\Phi(P_{k_l,i,l})$ defined as follows:

$$\Phi(P_{k_l,i,l}) = \pi_{k_l} \log_2 \left(1 + \beta \frac{P_{k_l,i,l} H_{k_l,i,l}}{I_{i,l} + \sigma_{i,l}^2} \right) - \sum_{m \neq l, m=1}^{L} \hat{c}_{i,m} P_{k_l,i,l} H_{k_l,i,m} - \lambda_{k_l} P_{k_l,i,l}$$

(11.25)

To maximize (11.24) for any given $\alpha_{k_l,i,l}$, a differentiation of (11.25) with respect to $P_{k_l,i,l}$ is performed and the result is set to 0. This yields

$$P_{k_l,i,l}^* = \left[\frac{\pi_{k_l}}{\ln(2) \left(\sum_{m \neq l, m=1}^{L} \hat{c}_{i,m} H_{k_l,i,m} + \lambda_{k_l} \right)} - \frac{I_{i,l} + \sigma_{i,l}^2}{\beta H_{k_l,i,l}} \right]^+$$

(11.26)

The second maximization can be treated by replacing, in (11.24), $P_{k_l,i,l}$ with the optimal value. This yields

$$D(\lambda) = \max_{\mathbf{A_l}} \sum_{k_l=1}^{K_l} \sum_{i=1}^{N} \left\{ \alpha_{k_l,i,l} \Phi(P_{k_l,i,l}^*) \right\} + \sum_{k_l=1}^{K_l} \lambda_{k_l} P_{k_l,\max}$$

(11.27)

By definition $\sum_{k_l=1}^{K_l} \alpha_{k_l,i,l} \leq 1$. Hence, the following can be noted:

$$\sum_{k_l=1}^{K_l} \alpha_{k_l,i,l} \Phi(P_{k_l,i,l}^*) \leq \max_{k_l} \Phi(P_{k_l,i,l}^*)$$

(11.28)

In fact, since time sharing is allowed, (11.28) corresponds to allocating each subcarrier i to the user k_l that maximizes $\Phi(P_{k_l,i,l}^*)$. This corresponds to a "winner takes all" on each subcarrier and the maximization over $\mathbf{A_l}$ reduces to the following rule

$$\alpha_{k_l,i,l}^* = \begin{cases} 1, & \text{if } k_l = k_{l,i}^* \\ 0, & \text{otherwise} \end{cases}$$

(11.29)

where

$$k_{l,i}^* = \arg\max_{k_l} \Phi(P_{k_l,i,l}^*)$$

(11.30)

Hence, although the problem was relaxed by allowing time sharing, the optimal solution reached consists of exclusive subcarrier allocation. The optimal power allocation is then $\tilde{P}_{k_l,i,l}^* = \alpha_{k_l,i,l}^* P_{k_l,i,l}^*$. This result in a multicell scenario is in accordance with the optimal subcarrier allocation in a single cell scenario, as derived in Chapter 3, and in Refs [32, 33, 95].

Finally, to complete the solution, the vector of geometric multipliers λ associated with the power constraint still needs to be determined. Determining the multipliers can be done using an iterative subgradient method with each iteration on the multiplier λ_{k_l} given by

$$\lambda_{k_l}^{n+1} = \left[\lambda_{k_l}^n - \delta_n G_{\lambda_{k_l}}^n\right]^+ \tag{11.31}$$

where the superscript n denotes the index of the iteration and G_λ^n denotes the subgradient, which is taken as:

$$G_{\lambda_{k_l}}^n = P_{k_l,\max} - \left\{\sum_{i=1}^N \alpha_{k_l,i,l}^* P_{k_l,i,l}^*\right\} \tag{11.32}$$

δ_n is the step size taken in the form $\delta_n = a/\sqrt{n}$ where a is a positive constant. This chosen step size is guaranteed to lead to convergence since it obeys the "nonsummable diminishing rule" [35]:

$$\lim_{n \to \infty} \delta_n = 0, \qquad \sum_{n=1}^{\infty} \delta_n = \infty. \tag{11.33}$$

11.3.3 Iterative Pricing Game

To solve (11.1), the BSs take turns in applying the solution of Section 11.3.2, for example, in a round-robin fashion. After the action taken by each BS, the interference prices in the network are updated based on the action taken. An iteration is complete after each BS has applied once the approach of Section 11.3.2. The process continues until convergence. The state of the network during iteration t with BSs 1 to l having performed their scheduling actions can be represented by the vector of pairs:
$\left\{(\mathbf{A}_1^t, \mathbf{P}_1^t), (\mathbf{A}_2^t, \mathbf{P}_2^t), ..., (\mathbf{A}_l^t, \mathbf{P}_l^t), (\mathbf{A}_{l+1}^{t-1}, \mathbf{P}_{l+1}^{t-1}), ..., (\mathbf{A}_L^{t-1}, \mathbf{P}_L^{t-1})\right\}$.

In other words, the action of BS l is determined by the actions of BSs 1 to $l-1$ at the current iteration, and the actions of BSs $l+1$ to L at the previous iteration. For example, each pricing term $\hat{c}_{i,m}$ in (11.15) is computed according to (11.10) and (11.12), where $P_{k_m,i,m}$ and $I_{i,m}$ obtained at the current iteration t are used if m is between 1 and $l-1$, but $P_{k_m,i,m}$ and $I_{i,m}$ obtained at the previous iteration $(t-1)$ are used if m is between $l+1$ and L. Whenever no confusion can occur, the index t will be dropped from the equations.

Now it will be shown that the pricing game described above converges to a Nash equilibrium. In fact, the pricing game in (11.15) follows a best response dynamic, since the approach of Section 11.3.2 leads to the best result for BS l, given the actions of the other BSs. Furthermore, the pricing game can easily be shown to be a potential game as defined in Ref. [173]. Since all potential games following a best response dynamic converge to a Nash equilibrium [172, 173], it can be concluded that the iterative pricing game converges to a Nash equilibrium. The details of the proof are presented below.

Each BS is a player in the iterative pricing game. The action a_l of BS l, selected from the set $\mathcal{A}_l$ of available actions, consists of determining $(\mathbf{A}_l, \mathbf{P}_l)$ or, equivalently, $\tilde{\mathbf{P}}_l$ such that each term in $\tilde{\mathbf{P}}_l$ satisfies $\tilde{P}_{k_l,i,l} = \alpha_{k_l,i,l} P_{k_l,i,l}$. Using standard game theory notation, when the interest is in the behavior of player l, the actions of all players are represented by $(a_l, \mathbf{a}_{-l})$, where $\mathbf{a}_{-l}$ is the vector of actions of all players except player l. The payoff of player l when the player actions are represented by $(a_l, \mathbf{a}_{-l})$ is denoted by $O_l(a_l, \mathbf{a}_{-l})$.

Definition 11.1. *A game is defined as a potential game if there exists a potential function X_{pot}, such that, for all players $l = 1, ..., L$ and for all actions $a_l, b_l \in \mathcal{A}_l$, the following applies:*

$$O_l(a_l, \mathbf{a}_{-l}) - O_l(b_l, \mathbf{a}_{-l}) = X_{\text{pot}}(a_l, \mathbf{a}_{-l}) - X_{\text{pot}}(b_l, \mathbf{a}_{-l}) \tag{11.34}$$

Theorem 11.1. *The pricing game in (11.15) follows a best response dynamic.*

Proof. It was shown that the approach of Section 11.3.2 leads to the best result for every BS, given the actions of the other BSs. In other words, the scheduling solution of Section 11.3.2 is optimal for BS l when the actions of BSs $1, 2, ..., l - 1, l + 1, ..., L$ are known at BS l. □

Theorem 11.2. *The pricing game defined in (11.15) is a potential game.*

Proof. In fact, the payoff of BS l represented by (11.18) can be equivalently expressed as

$$O_l(a_l, \mathbf{a}_{-l}) = \sum_{k_l=1}^{K_l} \sum_{i=1}^{N} \left\{ \pi_{k_l} \log_2 \left(1 + \beta \frac{\tilde{P}_{k_l,i,l} H_{k_l,i,l}}{I_{i,l} + \sigma_{i,l}^2} \right) - \hat{C}_{k_l,i} \tilde{P}_{k_l,i,l} \right\} \tag{11.35}$$

with $\hat{C}_{k_l,i} = \sum_{m \neq l, m=1}^{L} \hat{c}_{i,m} H_{k_l,i,m}$. Letting $\tilde{\mathbf{P}}_l$ correspond to action a_l and $\tilde{\mathbf{F}}_l$ correspond to action b_l, then

$$O_l(a_l, \mathbf{a}_{-l}) - O_l(b_l, \mathbf{a}_{-l}) = \sum_{k_l=1}^{K_l} \sum_{i=1}^{N} \left\{ \pi_{k_l} \log_2 \left(\frac{I_{i,l} + \sigma_{i,l}^2 + \beta \tilde{P}_{k_l,i,l} H_{k_l,i,l}}{I_{i,l} + \sigma_{i,l}^2 + \beta \tilde{F}_{k_l,i,l} H_{k_l,i,l}} \right) \right.$$

$$\left. - \hat{C}_{k_l,i}(\tilde{P}_{k_l,i,l} - \tilde{F}_{k_l,i,l}) \right\} \tag{11.36}$$

Considering the following potential function:

$$X_{\text{pot}}(a_l, \mathbf{a}_{-l}) = \sum_{l=1}^{L} \sum_{k_l=1}^{K_l} \sum_{i=1}^{N} \left\{ \pi_{k_l} \log_2 \left(I_{i,l} + \sigma_{i,l}^2 + \beta \tilde{P}_{k_l,i,l} H_{k_l,i,l} \right) - \hat{C}_{k_l,i} \tilde{P}_{k_l,i,l} \right\} \tag{11.37}$$

It can be clearly shown from (11.36) and (11.37) that (11.34) is verified. Hence the presented pricing game is a potential game. □

Theorem 11.3 is presented in Ref. [173] and is therefore repeated here without proof.

Theorem 11.3.   *All potential games following a best response dynamic converge to a Nash equilibrium.*

Using the results of Theorems 11.1, 11.2, and 11.3, it is straightforward to conclude that the iterative pricing game converges to a Nash equilibrium. However, this equilibrium is not shown to be unique. In fact, the uniqueness of the Nash equilibrium is derived in Ref. [53] for the single frequency case, but it is shown not to hold for the multiple frequency case. Hence, in such a complex OFDMA multicell scheduling scenario, the Nash equilibrium is not unique, and the convergence to a Pareto optimal solution is not guaranteed. Therefore, Section 11.4 presents an approach with central control that allows to reach a better equilibrium point than the iterative pricing game.

11.4   PRICING GAME WITH CENTRALIZED CONTROL

The Nash equilibrium does not always achieve the best social behavior (e.g., prisoner's dilemma [45]). The assumption that the CSI is known between a user and all the BSs in the network over all subcarriers is generally not achievable unless in the presence of central control in the network. If a CCU is present, it can intervene to enhance the sum-rate in the network regardless of the selfish behavior of BSs, since it knows the impact of each action taken on the total network rate. Thus, it can lead to a better solution than the Nash equilibrium obtained in Section 11.3. Consequently, in this section, the pricing game played between BSs is modified into a controlled pricing game. Hence, each BS plays a double game:

- A pricing game with other BSs. Given the prices set by the other BSs, each BS implements the solution derived in Section 11.3.2. This approach leads to the best achievable result for that BS, although it does not necessarily lead to an enhancement in the network sum-rate.

- A tit-for-tat game with the CCU. Sine the CCU is assumed to have central control over the resource allocation in the network, it accepts the action of BS l if it leads to an enhancement in the total network rate, and rejects that action if it leads to a deterioration in the network rate. Such a game is called tit-for-tat [45], since the CCU punishes BS l when it acts selfishly, and recompenses it in the opposite case.

Hence, the state of the network at any instant during iteration t can be represented by the vector of triplets: $\left\{ [(\mathbf{A}_1^t, \mathbf{P}_1^t), \chi_1^t], [(\mathbf{A}_2^t, \mathbf{P}_2^t), \chi_2^t], ..., [(\mathbf{A}_l^t, \mathbf{P}_l^t), \chi_l^t], \right.$ $\left. [(\mathbf{A}_{l+1}^{t-1}, \mathbf{P}_{l+1}^{t-1}), \chi_{l+1}^{t-1}], ..., [(\mathbf{A}_L^{t-1}, \mathbf{P}_L^{t-1}), \chi_L^{t-1}] \right\}$, where χ_l^t represents the action of the CCU, which consists of accepting the pair $(\mathbf{A}_l^t, \mathbf{P}_l^t)$ when it leads to an enhancement in the network, and of rejecting it by keeping $(\mathbf{A}_l^t, \mathbf{P}_l^t) = (\mathbf{A}_l^{t-1}, \mathbf{P}_l^{t-1})$ in the opposite case. The scenario of multicell scheduling with CCU is shown in Fig. 11.1.

It is trivial to show that the pricing game with central control converges to a solution that is better than (or, in the worst case, similar to) the solution of the pricing

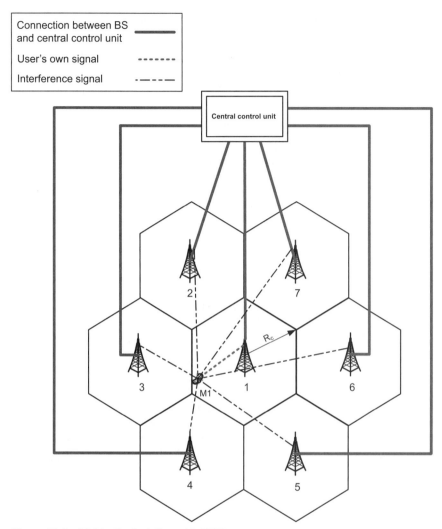

Figure 11.1 Multicell scheduling with CCU.

game without central control of Section 11.3 (in terms of network sum-rate). However, it cannot be shown that this solution is a unique Pareto optimal solution, for the same reasons outlined in Section 11.3.3 and in Ref. [53]. It should be noted that the pricing game with central control could lead to results that are unfair to certain BSs, that is, there might be certain cells where users are not allocated any subcarriers, due to the tit-for-tat strategy followed by the CCU. To deal with this problem, the CCU can accept all the actions of the BSs for a predefined number of iterations, then apply the tit-for-tat strategy for the remaining iterations until convergence. This allows a more fair competition between BSs during the first iterations.

11.4.1 Online Versus Offline Implementation

The pricing game without CCU can be applied online where each BS takes an action and the other BSs measure the interference and compute the prices accordingly. However, it is assumed that the CSI does not change during the iterative scheduling process [53, 172]. In the presence of a CCU, the game can be played offline since the CCU knows the CSI between all users and the BSs and can estimate the interference and determine the prices for each action taken by any BS. Then the CCU communicates the final scheduling decisions to all BSs after convergence.

In practical OFDMA systems, for example, WiMAX and LTE, a CCU is not generally present. Furthermore, scheduling cooperation between BSs is absent, for example, LTE [86], although it might be included in next-generation networks, for example, LTE-Advanced [89]. Hence, in the following section, a scheduling approach is presented. The presented approach can be applied online without iterations, does not assume that the CSI of the users is known for all BSs, and can use pricing-based power control without BS cooperation.

11.5 SUBOPTIMAL SCHEDULING SCHEME USING PRICING-BASED POWER CONTROL

In this section, a suboptimal scheduling approach is presented. Each BS performs resource allocation in its own cell using a low complexity scheduling algorithm. In the presented approach, the CSI between users and their serving BS are known to the BS. However, the CSI between the users and other BSs are not known. These assumptions are more practical than the stringent requirements of the iterative game theoretical solution. The approach relies on pricing-based power control, and can be applied with and without BS cooperation. Hence, two scenarios are investigated:

- *Scenario Without BS Cooperation*: In this case, there is no exchange of information between BSs. Each BS implements a power control scheme that does not require coordination with other BS. The noncooperative model is shown in Fig. 11.2.

- *Scenario with BS Cooperation*: In this case, the BSs cooperate by exchanging interference information. Since the highest interference is received from the direct neighbors, the exchange of information of a given BS is limited to its six surrounding cells in order to avoid excessive complexity in the system. The cooperative model is shown in Fig. 11.3.

11.5.1 Utility Functions

Utility maximization is considered, where $U_{k_l}(R_{k_l})$ is the utility of user k_l and R_{k_l} is its rate given by

$$R_{k_l}\left(\mathbf{P}_{\mathbf{k_l}}, \mathcal{I}_{\text{sub},k_l}\right) = \sum_{i=1}^{N} \alpha_{k_l,i,l} \frac{B}{N} \log_2\left(1 + \beta\gamma_{k_l,i,l}\right) \tag{11.38}$$

where $\mathbf{P}_{\mathbf{k_l}}$ represents a vector whose elements are $P_{k_l,i,l}$, the transmitted power by user k_l on each subcarrier. The total bandwidth is represented by B. Consequently,

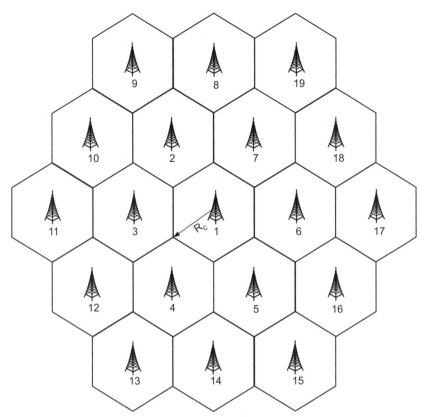

Figure 11.2 System model: noncooperative scenario.

B/N is the bandwidth per subcarrier. In addition, $\gamma_{k_l,i,l}$ is the signal to interference plus noise ratio (SINR) of user k_l over subcarrier i in cell l. It is given by

$$\gamma_{k_l,i,l} = \frac{P_{k_l,i,l} H_{k_l,i,l}}{\displaystyle\sum_{j\neq l, j=1}^{L} \sum_{k_j=1}^{K_j} \alpha_{k_j,i,j} P_{k_j,i,j} H_{k_j,i,l} + \sigma_{i,l}^2} \tag{11.39}$$

Letting the utility equal to the rate leads to a greedy maximization of the sum-rate of the cell, whereas setting the utility to the rate divided by the average data rate achieved by the user at previous TTIs corresponds to proportional fairness [48]. Both greedy and proportional fair (PF) utilities are separable across the subcarriers, that is, they satisfy $U_{k_l}(R_{k_l}) = \sum_{i=1}^{N} U_{k_l,i}(R_{k_l,i})$, where $U_{k_l,i}$ is the utility of user k_l over subcarrier i and $R_{k_l,i}$ is the rate of user k_l over subcarrier i.

To implement power control with greedy scheduling, the following utility is used for user k_l over subcarrier i instead of the rate:

$$U_{k_l,i}\left(R_{k_l,i}(P_{k_l,i,l}, \mathcal{I}_{\text{sub},k_l})\right) = \alpha_{k_l,i,l}\left(\log_2\left(1 + \beta\gamma_{k_l,i,l}\right) - C_{i,l}P_{k_l,i,l}\right) \tag{11.40}$$

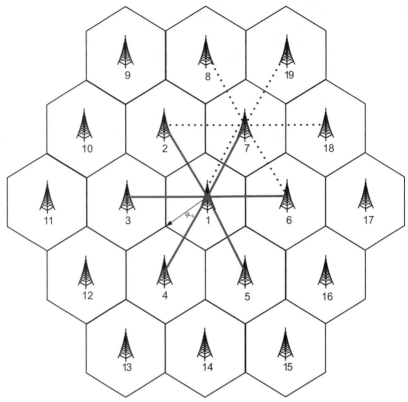

Figure 11.3 System model: cooperative scenario. The solid lines represent exchange of information with BS1 whereas the dotted lines represent exchange of information with BS7.

where $C_{i,l}$ is the cost of transmission over subcarrier i in cell l. Setting the value of $C_{i,l}$ will be discussed in Section 11.5.2.

Greedy scheduling maximizes the sum-rate of the cell but is unfair to edge users since most resources are allocated to users near the cell center, whereas edge users will generally suffer from starvation. To solve this problem, utility functions providing proportional fairness are desired. To use PF scheduling with power control, (11.40) can be extended to reflect proportional fairness as follows:

$$U_{k_l,i}\left(R_{k_l,i}(P_{k_l,i,l}, \mathcal{I}_{\text{sub},k_l})\right) = \frac{\alpha_{k_l,i,l}\left(\log_2\left(1 + \beta\gamma_{k_l,i,l}\right) - C_{i,l}P_{k_l,i,l}\right)}{D_{k_l,\text{tot}}} \quad (11.41)$$

where $D_{k_l,\text{tot}} = \sum_{t=1}^{n} R_{k_l}^{(t)}/n$ is the average rate achieved by user k_l over the last n TTIs. Setting the utility to the rate divided by the average data rate achieved by the user at previous TTIs also corresponds to proportional fairness [48], as expressed by (7.15) presented in Section 7.4, and leads to a separable utility, which facilitates the implementation of pricing-based power control [53]. The utility in (11.41) corresponds to proportional fairness in time and frequency (PFTF), since it involves frequency scheduling on a per subcarrier basis while taking into account the achieved data rate at previous TTIs. It should be noted that the PFF scheduling utilities described in Chapter 7 are not separable across subcarriers.

11.5.2 Setting the Prices in the Power Control Scheme

The cost of transmission on subcarrier i in cell l is given by $C_{i,l} = S(\sum_{j \neq l, j=1}^{L} \bar{c}_{i,j})$, with $\bar{c}_{i,j}$ taking a value in the interval [0 1] and S a scaling factor to increase the price beyond [0 1]. It is assumed to have a constant value used by all BSs. In the BS collaboration scenario, $\bar{c}_{i,j}$ represents the price imposed by cell j on the transmission over subcarrier i in neighboring cells. It is given by

$$\bar{c}_{i,j} = \min \left(\frac{I_{i,j}}{P_{\text{ref}}}, 1 \right) \tag{11.42}$$

where $I_{i,j}$ is the interference on subcarrier i measured at BS j, and P_{ref} is a power threshold above which the price is set to one. It indicates the interference level above which the BS sends the highest price to its neighbors. The expression of the interference is given by (11.5). It should be noted that BS j is not able to know all the individual terms in the double summation of (11.5). However, BS j can measure the total interference power, that is, the received power on subcarrier i when no transmissions are originating from cell j on subcarrier i. In the absence of BS collaboration, all BSs in the network set $C_{i,l} = S$ for all i. The power on each subcarrier when power control is used can be obtained by setting the derivative of (11.40) to zero. This yields

$$P_{k_l,i,l} = \left[\frac{1}{\ln(2)C_{i,l}} - \frac{1}{\beta H_{k_l,i,l}/\sigma_i^2} \right]^+ \tag{11.43}$$

The power in (11.43) has the desirable property of increasing when the channel gain $H_{k_l,i,l}$ increases and decreasing when the interference price $C_{i,l}$ increases, thus having a similar behavior to the power expressions in (11.14) and (11.26). However, the expression in (11.43) does not include Lagrangian parameters, and hence does not involve subgradient iterations to compute the Lagrangians as in (11.14) and (11.26).

11.5.3 Scheduling Algorithm

In this section, Algorithm 11.1, which is a low complexity scheduling algorithm that can be used with various utility functions, is presented. The algorithm consists of allocating subcarrier i to user k_l in a way to maximize the difference

$$\Lambda_{k_l,i} = U_{k_l}(R_{k_l}\left(\mathbf{P_{k_l}}, \mathcal{I}_{\text{sub},k_l} \cup \{i\}\right)) - U_{k_l}\left(R_{k_l}(\mathbf{P_{k_l}}, \mathcal{I}_{\text{sub},k_l})\right) \tag{11.44}$$

where the marginal utility, $\Lambda_{k_l,i}$, represents the gain in the utility function U_{k_l} when subcarrier i is allocated to user k_l, compared to the utility of user k_l before the allocation of i.

Algorithm 11.1

Multicell Scheduling Algorithm with Linear Complexity
for k_l such that $1 \leq k_l \leq K_l$
    $\mathcal{I}_{\mathrm{sub},k_l} \leftarrow \emptyset$
end for
for i such that $1 \leq i \leq N$
    for all k such that $1 \leq k_l \leq K_l$
        Compute $U_{k_l}(P_{k_l}, \mathcal{I}_{\mathrm{sub},k_l} \cup \{i\})$ using (11.43) derived in Section 11.5.2 for the power allocated on subcarrier i
        if $\mathcal{I}_{\mathrm{sub},k_l} \neq \emptyset$
            Compute $U_{k_l}(P_{k_l}, \mathcal{I}_{\mathrm{sub},k_l})$
        else
            $U_{k_l}(P_{k_l}, \mathcal{I}_{\mathrm{sub},k_l}) \leftarrow 0$
        end if
        $\Lambda_{k_l,i} \leftarrow U_{k_l}(P_{k_l}, \mathcal{I}_{\mathrm{sub},k_l} \cup \{i\}) - U_{k_l}(P_{k_l}, \mathcal{I}_{\mathrm{sub},k_l})$
    end for
    $k_l^* \leftarrow \arg \max_{k_l} \Lambda_{k_l,i}$
    if $\Lambda_{k_l^*,i} > 0$
        Subcarrier Allocation:
        $\mathcal{I}_{\mathrm{sub},k_l^*} \leftarrow \mathcal{I}_{\mathrm{sub},k_l^*} \cup \{i\}$
        Power Allocation:
        Allocate power on subcarrier i according to (11.43) derived in Section 11.5.2. If, for certain users, the result is such that $\sum_{i=1}^{N} P_{k_l,i,l} > P_{k_l,\max}$, enforce the power constraint by setting $P_{k_l,i,l} = P_{k_l,i,l} \cdot P_{k_l,\max} / \left(\sum_{i=1}^{N} P_{k_l,i,l} \right)$.
    end if
end for

The presented algorithm allocates each subcarrier after performing a linear search on the users in order to find the user that maximizes the marginal utility. Consequently, the total complexity of the algorithm is $\mathcal{O}(NK_l)$, that is, the algorithm has linear complexity in the number of users and in the number of subcarriers.

11.6 SUBOPTIMAL SCHEDULING SCHEME USING PROBABILISTIC TRANSMISSION

In this section, an intercell interference avoidance technique based on probabilistic transmission without BS cooperation is presented. Each BS measures the received interference power on each subcarrier. Then, each BS l computes a price $\bar{c}_{i,l}$ on each subcarrier i based on the measured interference, according to (11.42). Since $\bar{c}_{i,l}$ is selected to take a value in the interval $[0\ 1]$, it can be used as a probability measure. Hence, at each scheduling interval, the BS in cell l decides to stop using subcarrier i with probability $\bar{c}_{i,l}$, or continues using it with probability $(1 - \bar{c}_{i,l})$, that is, when the interference on a given subcarrier increases, its chances of being shut down also increase.

With this probabilistic scheduling approach, Algorithm 7.2 is applied on the subcarriers that are still "on." Hence, since pricing-based power control is not applied, the separability of the utility over the subcarriers is not required. Consequently, the logarithmic utility can be used for PF scheduling. Considering the average user rate in the logarithmic utility leads to PFTF scheduling, whereas considering the rate at the current TTI only leads to PFF scheduling, as expressed by (7.6) and the discussion in Section 7.4.

With the probabilistic scheduling approach, each user is assumed to transmit at the maximum power ($P_{k_l} = P_{k_l,\max}$), and the power is assumed to be subdivided equally among all the subcarriers allocated to that user. Hence, the approach is not based on power control, but rather on shutting down the subcarriers subjected to relatively high interference, and concentrating the power on the remaining subcarriers.

11.7 RESULTS AND DISCUSSION

11.7.1 Simulation Model

The simulation model consists of a wireless OFDMA network containing 19 cells. Users are uniformly distributed in the network. A wrap-around technique is applied in the simulations in order to remove edge effects [174]. The cell radius is set to $R_c = 500$ m from the BS. The total bandwidth is $B = 5$ MHz subdivided into $N = 16$ subcarriers. Equal user weights are assumed. A target BER of 10^{-6} is considered. The maximum user transmit power is assumed to be 125 mW. The duration of one TTI is considered to be 1 ms. The channel gain over subcarrier i between user k_l in cell l and BS j is given by

$$H_{k_l,i,j,\mathrm{dB}} = (-\kappa - \upsilon \log_{10} d_{k_l,j}) - \xi_{k_l,i,j} + 10 \log_{10} F_{k_l,i,j} \qquad (11.45)$$

In (11.45), the first factor captures propagation loss, with κ a constant chosen to be 128.1 dB, $d_{k_l,j}$ the distance in km from user k_l to BS j, and υ the path loss exponent, which is set to a value of 3.76. The second factor, $\xi_{k_l,i,j}$, captures log-normal shadowing with an 8 dB standard deviation, whereas the last factor, $F_{k_l,i,j}$, corresponds to Rayleigh fading with a Rayleigh parameter b such that $E[b^2] = 1$. The threshold power P_{ref} is considered to be the power received from a user transmitting at its maximum power from a reference distance d_{ref} from the BS, when only pathloss is present (shadowing and fading are not considered in P_{ref} in order for it to have a constant value since it is a power threshold). In the suboptimal pricing-based power control schemes, the values $S = 10$ and $d_{\mathrm{ref}} = 0.1 R_c$ are used. It is assumed that the channel gain (11.45) remains approximately constant during the duration of 10 TTIs.

11.7.2 Comparison of the Pricing-Based Power Control Schemes

The network sum-rate results are shown in Fig. 11.4, and the total transmit power in the network is shown in Fig. 11.5, for the case of 10 users per cell. The convergence

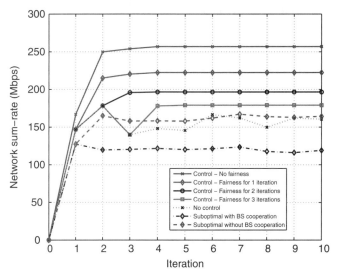

Figure 11.4 Network sum-rate variation versus the number of iterations.

of most schemes occurs after two to three iterations. The pricing game with CCU achieves a considerably higher rate with considerably less transmit power than the pricing game without CCU. The results of the pricing game with the CCU accepting all the actions of the BSs for the first one, two, and three iterations coincide with the first one, two, and three iterations, respectively, of the pricing game without CCU. The suboptimal scheduling scheme without BS cooperation achieves a rate comparable to

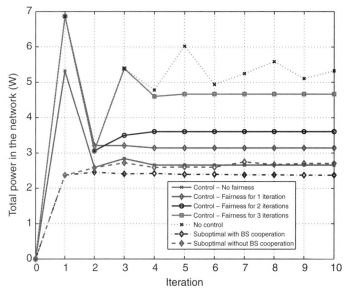

Figure 11.5 Network total transmit power variation versus the number of iterations.

the pricing game without CCU, with the benefit of a considerably reduced transmit power. This result is important, since it shows that efficient interference mitigation can be achieved using low complexity scheduling techniques that do not involve sub-gradient iterations, do not require BS collaboration, and do not assume that the BSs are aware of the CSI of users in other cells. In addition, the suboptimal scheduling scheme without BS cooperation outperforms the case with BS cooperation. Although this might seem as an unexpected result, it can be explained by the fact that exchanging total interference information between neighboring BSs according to (11.42) does not represent accurate pricing toward each individual BS, since each BS is contributing differently in the total interference. In fact, with the pricing scheme of (11.42), BS l penalizes all its surrounding BSs equally using the total received interference, regardless of the interference caused by each BS. This leads to reduced penalization in some cases and excessive penalization in other cases. However, this pricing scheme is used since a BS cannot distinguish the interference caused by each neighboring BS. Conversely, in the scenario without cooperation, setting a fixed price for all BSs over all subcarriers contributes in reducing the overall interference level in the network, thus leading to enhanced results. Tables 11.1 and 11.2 show the sum-rate and total transmit power, respectively, in each of the 19 cells at the end of the 10th iteration with the different investigated pricing-based techniques. Comparing the first two columns in Tables 11.1 and 11.2, it can be seen that the enhancement due to the presence of

TABLE 11.1   Sum-Rate (Mbps) in Each Cell at the 10th Iteration

Cell no.	No control	Control with no fairness	Control-fairness for 1 it.	Control-fairness for 2 it.	Control-fairness for 3 it.	Subopt. coop	Subopt. no coop
1	3.49	0	1.46	0.66	9.27	3.05	2.53
2	6.05	7.28	6.08	4.70	5.91	7.17	7.51
3	14.71	40.68	38.47	26.14	17.67	13.72	14.34
4	13.44	25.52	15.86	11.66	11.05	6.09	10.85
5	3.28	0.86	0	6.23	0.96	0.28	2.42
6	12.01	29.32	24.10	11.93	11.13	11.39	15.84
7	13.02	7.04	5.93	9.66	9.28	8.89	11.78
8	6.99	10.21	20.13	26.72	14.56	6.58	5.95
9	4.82	7.40	0.46	3.20	7.48	1.78	7.74
10	6.48	0	0.77	1.88	8.13	1.29	2.27
11	6.01	0.38	4.70	6.07	2.31	6.08	7.69
12	0.95	0.05	0	4.88	4.43	0.86	1.76
13	11.78	44.25	14.36	11.32	14.64	11.32	13.66
14	7.10	0	8.58	12.15	9.54	2.59	4.66
15	3.54	2.82	1.73	0.65	6.91	2.90	6.09
16	22.55	54.04	46.33	34.16	26.92	23.96	33.46
17	6.01	0	6.94	6.47	2.27	1.85	2.92
18	4.53	0	8.54	6.16	5.72	0.56	3.43
19	13.73	27.01	17.1	11.90	10.88	8.65	9.39

TABLE 11.2 Sum of Transmit Power (W) in Each Cell at the 10th Iteration

Cell no.	No control	Control with no fairness	Control-fairness for 1 it.	Control-fairness for 2 it.	Control-fairness for 3 it.	Subopt. coop	Subopt. no coop
1	0.29	0	0.062	0.062	0.15	0.125	0.125
2	0.125	0.067	0.066	0.066	0.166	0.125	0.110
3	0.25	0.5	0.5	0.375	0.25	0.125	0.129
4	0.25	0.25	0.131	0.097	0.25	0.125	0.130
5	0.453	0.121	0	0.179	0.115	0.125	0.210
6	0.25	0.375	0.375	0.156	0.251	0.125	0.119
7	0.25	0.021	0.020	0.038	0.25	0.125	0.125
8	0.375	0.258	0.308	0.375	0.375	0.125	0.158
9	0.25	0.177	0.030	0.037	0.281	0.125	0.140
10	0.484	0	0.034	0.078	0.625	0.125	0.277
11	0.125	0.001	0.125	0.125	0.066	0.125	0.125
12	0.141	0.016	0	0.125	0.125	0.125	0.159
13	0.25	0.25	0.112	0.137	0.25	0.125	0.138
14	0.449	0	0.135	0.5	0.5	0.125	0.138
15	0.125	0.125	0.125	0.109	0.125	0.125	0.116
16	0.25	0.25	0.25	0.25	0.026	0.125	0.127
17	0.496	0	0.25	0.269	0.25	0.125	0.194
18	0.262	0	0.369	0.375	0.361	0.125	0.080
19	0.25	0.25	0.25	0.25	0.25	0.125	0.115

the CCU comes at the expense of fairness, since several cells in the second column achieve a zero rate (cells 1, 10, 14, 17, and 18) whereas others achieve a very limited rate (cells 5, 11, and 12). As the CCU does not intervene for one, two, and three iterations, the number of cells with reduced rate decreases as can be seen in the third, fourth, and fifth columns, respectively.

Varying the number of users and considering the output results at the tenth iteration, Fig. 11.6 shows the network sum-rate and Fig. 11.7 shows the network efficiency, defined as the number of rate bits achieved by unit energy transmitted [172], that is,

$$\eta = \left(\sum_{l=1}^{L} \sum_{k_l=1}^{K_l} R_{k_l} \right) \Big/ \left(\sum_{l=1}^{L} \sum_{k_l=1}^{K_l} P_{k_l} \right) \qquad (11.46)$$

Fig. 11.6 shows that the pricing-based scheme without central control achieves a sum-rate comparable to the suboptimal scheme without BS cooperation and better than the suboptimal scheme with BS cooperation. However, it can be seen from Fig. 11.7 that scheduling with centralized control and the suboptimal schemes with and without BS cooperation are more efficient than the pricing-based scheme without central control. Hence, although the pricing-based scheme without CCU led to

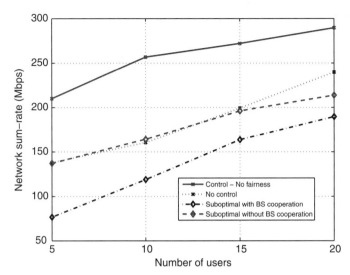

Figure 11.6   Network sum-rate versus the number for users: results at the 10th iteration.

better sum-rate results than the suboptimal scheme with BS cooperation, this comes at the expense of considerably higher transmit power, as can be concluded from Fig. 11.7. It should be noted that showing the sum-rate and efficiency results makes the power results redundant, since the transmit power can be deduced from the rate and efficiency plots.

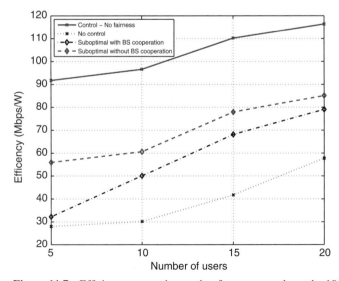

Figure 11.7   Efficiency versus the number for users: results at the 10th iteration.

11.7.3 Results of the Suboptimal Pricing-Based Power Control Schemes

In this section, the practical scenario of online scheduling with power control is considered, that is, instantaneous resource allocation is performed at all BSs simultaneously without having the BSs take turns in a round-robin fashion, using the methods described in Section 11.5. The performance of the suboptimal power control schemes with and without BS cooperation is compared to the case of scheduling without applying any interference mitigation techniques. Fig. 11.8 shows the average cell rate results for greedy and PFTF utilities, and the efficiency results are shown in Fig. 11.9. Clearly, interference mitigation using pricing-based power control led to considerable enhancements compared to scheduling without interference mitigation. As the number of users increases, the efficiency values of the cooperative and noncooperative schemes converge to approximately similar results, for both greedy and PFTF scheduling. Greedy scheduling leads to a higher rate and efficiency than PFTF. This is expected, since the fairness of PFTF comes at the expense of reduced cell rate for the same transmit power. Fig. 11.9 also shows that the results with power control are an order of magnitude more efficient than those obtained by scheduling without power control, for both greedy and PFTF scheduling.

It should be noted that the sum-rate results do not correspond to the PFTF utility, since with PF scheduling a logarithmic utility is used (although (11.41) is used to approximate the logarithmic utility during the scheduling process, the sum of the logarithms of the rates achieved after the application of Algorithm 11.1 can still be computed). Fig. 11.10 shows the average cell sum-utility results, with the user utility given by $U_{k_l} = \ln \left(\sum_t R_{k_l}^{(t)} \right)$, where the summation is taken over 300 TTIs.

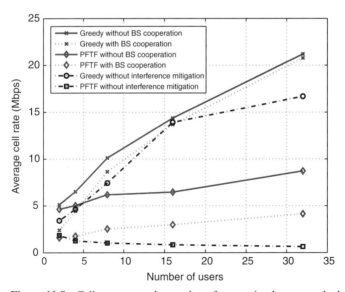

Figure 11.8 Cell rate versus the number of users: simultaneous scheduling without iterations using the suboptimal pricing-based power control schemes.

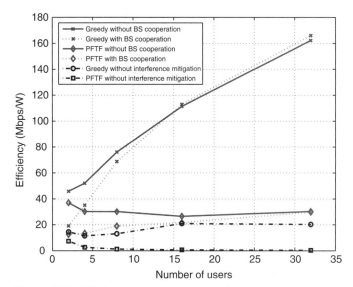

Figure 11.9   Efficiency versus the number of users: simultaneous scheduling without iterations using the suboptimal pricing-based power control schemes.

It should be noted that, in the simulations, when a user has zero rate, and to avoid dealing with $-\infty$ values, the rate of that user is set to a very low value ϵ instead, with $\epsilon = 2.2 \times 10^{-16}$ in Matlab for example. This explains the decrease in the utility function and the negative values with the greedy scheduling scenario. This decrease

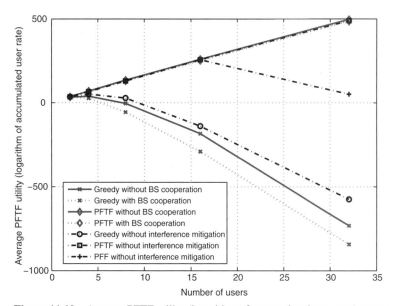

Figure 11.10   Average PFTF utility (logarithm of accumulated user rate) versus the number of users: simultaneous scheduling without iterations using the suboptimal pricing-based power control schemes.

TABLE 11.3   Average Number of Allocated Subcarriers per Cell per TTI with the Pricing-Based Suboptimal Power Control Schemes

	No. of Users				
	2	4	8	16	32
Greedy-no PC	15.59	16	16	16	16
Greedy-BS coop.	13.82	14.98	15.58	15.73	15.98
Greedy-w/o BS coop.	8.92	10.74	12.86	14.13	15.55
PFTF-no PC	15.58	16	16	16	16
PFTF-BS coop.	13.49	14.10	14.47	14.84	15.24
PFTF-w/o BS coop.	8.91	10.11	11.56	12.42	14.09

does not occur for PFTF, since it can ensure fairness over the time dimension that provides an additional degree of freedom.

Table 11.3 shows the average number of allocated subcarriers per cell per TTI. By comparing Table 11.3 to the results of Figs 11.8 and 11.9, it can be seen that the best results were obtained when the number of scheduled subcarriers is the least, both for the greedy and PFTF scheduling utilities. This is explained by the fact that the presented pricing-based power control scheme without BS cooperation was able to enhance the OFDMA network performance by turning off the subcarriers subjected to relatively high interference.

11.7.4   Results of the Suboptimal Probabilistic Scheduling Scheme

In this section, simulation results are presented for the probabilistic scheduling approach of Section 11.6. At each TTI, the BSs measure the interference and perform scheduling in each cell based on the presented probabilistic interference avoidance method. Results are presented with PFF and PFTF utilities. Results with the greedy sum-rate utility were generated, but the probabilistic scheduling scheme led to results comparable to the case without interference mitigation, and hence no improvement was obtained. Therefore, results with greedy scheduling are not presented.

Fig. 11.11 shows the average cell rate results for PFF and PFTF with and without interference mitigation. Scenarios with $d_{\text{ref}} = 0.1 R_c$ and $0.2 R_c$ are considered. Clearly, the probabilistic interference mitigation scheme led to considerable enhancements compared to scheduling without interference mitigation, especially with PFTF. The PFF curves do not converge as the number of users increases, conversely to PFTF curves. This behavior is explained by Theorem 7.1 presented in Chapter 7, since when $K_l > N$, PFF allocates one subcarrier to each of the N users having the best channel conditions.

Fig. 11.12 shows the efficiency results. As the number of users increases, the efficiency decreases, as expected, due to the increase in intercell interference. In

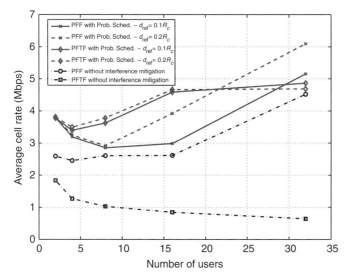

Figure 11.11 Cell rate versus the number of users: simultaneous scheduling without iterations using the suboptimal probabilistic scheduling scheme. This figure is adapted from E. Yaacoub and Z. Dawy, "Proportional Fair Scheduling with Probabilistic Interference Avoidance in the Uplink of Multicell OFDMA Systems", IEEE GlobeCom Workshops (MCECN), Miami FL, USA, December 2010 ©2010 IEEE.

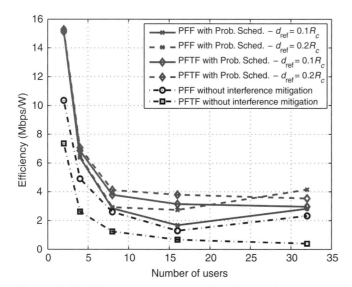

Figure 11.12 Efficiency versus the number of users: simultaneous scheduling without iterations using the suboptimal probabilistic scheduling scheme. This figure is adapted from E. Yaacoub and Z. Dawy, "Proportional Fair Scheduling with Probabilistic Interference Avoidance in the Uplink of Multicell OFDMA Systems", IEEE GlobeCom Workshops (MCECN), Miami FL, USA, December 2010 ©2010 IEEE.

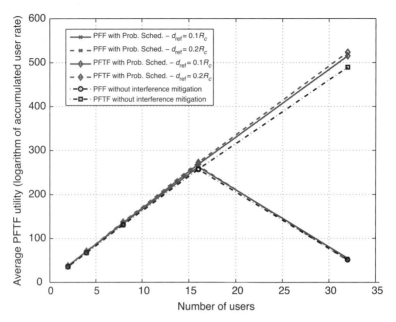

Figure 11.13 Average PFTF utility (logarithm of accumulated user rate) versus the number of users: simultaneous scheduling without iterations using the suboptimal probabilistic scheduling scheme. This figure is adapted from E. Yaacoub and Z. Dawy, "Proportional Fair Scheduling with Probabilistic Interference Avoidance in the Uplink of Multicell OFDMA Systems", IEEE GlobeCom Workshops (MCECN), Miami FL, USA, December 2010 ©2010 IEEE.

addition, the case $d_{\mathrm{ref}} = 0.2R_c$ leads to more efficient results than $d_{\mathrm{ref}} = 0.1R_c$. Furthermore, the presented scheme leads to around four times more efficient results with PFTF and two times more efficient results with PFF and $d_{\mathrm{ref}} = 0.2R_c$, compared to the case without probabilistic interference avoidance.

It should be noted that the sum-rate results do not accurately reflect utility enhancements, since with PF scheduling a logarithmic utility is used. Fig. 11.13 shows the average cell sum-utility results, with the user utility given by $U_{k_l} = \ln\left(\sum_t R_{k_l}^{(t)}\right)$, where the summation is taken over $n = 300$ TTIs as in Section 11.7.3, and a similar value of ϵ is used. This explains the linear decrease in the utility function with the PFF scenario in Fig. 11.13 when K_l becomes greater than N. This decrease does not occur for PFTF, since it can ensure fairness over the time dimension which provides an additional degree of freedom. By comparing Fig. 11.13 to Fig. 11.10, it can be seen that the probabilistic scheduling scheme led to better PFTF utility results than the pricing-based schemes. However, by comparing Fig. 11.12 to Fig. 11.9, it can be seen that the pricing-based schemes are considerably more efficient, since the probabilistic scheduling approach applies maximum power transmission.

Table 11.4 shows the average number of allocated subcarriers per cell per TTI. By comparing Table 11.4 to the results of Figs 11.11–11.13, it can be seen that the probabilistic interference avoidance scheme was able to enhance the OFDMA network

TABLE 11.4   Average Number of Allocated Subcarriers per Cell per TTI with the Probabilistic Scheduling Scheme

	No. of users				
	2	4	8	16	32
PFF without interference avoidance	15.59	16	16	16	16
PFF with Prob. Sched. $d_{ref} = 0.1 R_c$	10.40	12.11	13.72	14.18	14.58
PFF with Prob. Sched. $d_{ref} = 0.2 R_c$	9.86	10.97	11.54	11.42	11.75
PFTF without interference avoidance	15.58	16	16	16	16
PFTF with Prob. Sched. $d_{ref} = 0.1 R_c$	10.45	12.31	14.02	14.40	14.34
PFTF with Prob. Sched. $d_{ref} = 0.2 R_c$	9.89	11.12	11.85	11.63	11.46

This table is reprinted, with permission from IEEE, from E. Yaacoub and Z. Dawy, "Proportional Fair Scheduling with Probabilistic Interference Avoidance in the Uplink of Multicell OFDMA Systems", IEEE GlobeCom Workshops (MCECN), Miami FL, USA, December 2010 ©2010 IEEE.

performance by turning off the subcarriers subjected to relatively high interference and concentrating the transmit power of the different users on the least interfered subcarriers, in conjunction with an efficient subcarrier allocation scheme.

11.8   SUMMARY

The problem of network sum-rate maximization in the uplink of OFDMA systems was formulated and solved based on a pricing game between BSs. The solution was enhanced when a CCU is assumed to play a tit-for-tat game with each BS. Suboptimal scheduling schemes with and without BS cooperation were compared to the iterative solutions based on game theory. The suboptimal pricing-based scheduling approach without BS cooperation was shown to outperform the collaborative case based on exchanging total interference information. In addition, it performed better than the iterative pricing game solution without CCU. A probabilistic scheduling approach was presented for the case without BS cooperation. Compared to the case without interference mitigation, the probabilistic scheduling scheme led to better results by successfully turning off the subcarriers subjected to high interference.

CHAPTER 12

DISTRIBUTED MULTICELL SCHEDULING WITH INTERFERENCE MITIGATION

In Chapter 11, centralized multicell scheduling was investigated. In this chapter, distributed resource allocation in a multicell scenario is investigated while using, in each cell, the distributed scheduling approach studied in Chapter 9. This chapter is organized as follows. Section 12.1 presents the background information and lists the topics of interest studied in this chapter. The system model is presented in Section 12.2. The intracell cooperative approach is described in Section 12.3. The intercell interference mitigation/avoidance schemes with and without CCD collaboration are discussed in Section 12.4. The simulation results are presented and analyzed in Section 12.5. Practical implementation aspects are discussed in Section 12.6. The chapter summary is presented in Section 12.7.

12.1 BACKGROUND

In Chapter 9, a single cell distributed scheduling scheme was presented. It was based on the cooperation between users within the cell. In Chapter 11, centralized scheduling in a multicell scenario was investigated. Practical scheduling schemes with and without BS collaboration were studied. In this chapter, distributed uplink scheduling in multicell OFDMA systems is investigated. The investigated scenario is based on the collaboration in two levels. The first level is an intracell level, where the users in a each cell cooperate by exchanging CSI and implementing a distributed scheduling algorithm, as described in Chapter 9. The second level is an intercell level where the CCDs in each cell cooperate by exchanging interference information. A pricing-based power control scheme using the exchanged information is presented, in addition to a transparent scheme where users are unaware of the exchanged information. Furthermore, a probabilistic interference avoidance technique is presented in the case where CCDs do not cooperate. The technique is based on shutting down each subcarrier with a probability that increases with the received interference level on that subcarrier.

A distributed interference management approach is proposed in Refs [105, 171]. The approach is based on a busy burst transmission on the subcarriers

Resource Allocation in Uplink OFDMA Wireless Systems: Optimal Solutions and Practical Implementations,
Elias E. Yaacoub and Zaher Dawy.
© 2012 by the Institute of Electrical and Electronics Engineers, Inc. Published 2012 by John Wiley & Sons, Inc.

occupied by a user. Other users hearing the busy bursts above a certain threshold will not transmit on these subcarriers, which creates a sort of protection zone around the user transmitting the bursts, thus reducing intercell interference. This method is particularly useful for cell edge users, and it represents another example for distributed intercell interference management than the methods described in this chapter. Probabilistic interference avoidance is used in 3G CDMA cellular systems, for example, cdma2000 1xEV-DO, in order to mitigate intracell interference when distributed scheduling is applied [64, 65]. The uplink traffic channel rate control algorithm for cdma2000 is presented in Ref. [153], where transition from one rate to another is performed by mobile devices in a probabilistic manner. In this chapter, probabilistic transmission is used for intercell interference avoidance in the case of distributed scheduling in OFDMA uplink without intercell collaboration.

The main topics of interest presented in this chapter can be summarized as follows:

- Extending the distributed scheduling approach described in Chapter 9 to a multicell scenario in the presence of intercell interference.

- Applying collaborative intercell interference mitigation via pricing-based power control. This approach includes exclusive subcarrier allocation, considers separate prices on each subcarrier, and is not based on lengthy iterative techniques.

- Presenting a transparent pricing scheme for collaborative intercell interference mitigation. The scheme consists of elegantly embedding the pricing information in the pilot signals transmitted by the CCD. This will lead to an "artificial" variation of the CSI measured by the users. Users perform distributed intracell scheduling based on quantized CSI without being aware of the exchanged pricing information between the CCDs. The transparent scheme includes subcarrier allocation, does not rely on explicit power control, considers separate prices on each subcarrier, is not based on iterative techniques, and is completely transparent to the users.

- Implementing a probabilistic interference avoidance technique in the case where CCD collaboration is not available. The technique is based on shutting down each subcarrier with a probability that increases with the received interference level on that subcarrier. This technique is also transparent to the users who implement distributed scheduling with quantized CSI using only the subcarriers that are "on", that is, the subcarriers on which they received a pilot signal from the CCD.

- Discussing the applicability of the presented interference mitigation/avoidance techniques in different practical scenarios.

12.2   SYSTEM MODEL

The system studied consists of a set of CCDs, each covering a small area. The reduced coverage area is assumed because the users are required to be in a relatively close

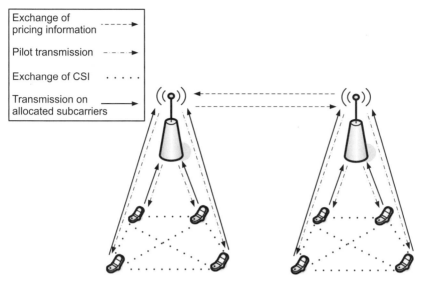

Figure 12.1 System model in the case of two cooperating CCDs. This figure is reprinted, with permission from IEEE, from E. Yaacoub and Z. Dawy, "A Transparent Pricing Scheme for Interference Mitigation in Uplink OFDMA with Collaborative Distributed Scheduling", IEEE International Conference on Telecommunications (ICT 2010), Doha, Qatar, April 2010 ©2010 IEEE.

proximity so that they can collaborate by exchanging information. The CCD can have the same practical implementations discussed in Section 9.2.1. Two scenarios are investigated:

1. *Scenario with CCD Cooperation*: In this case, the model consists of intracell cooperation between users by exchanging CSI information and intercell cooperation between CCDs by exchanging interference information. An example with two CCDs is shown in Fig. 12.1.

2. *Scenario Without CCD Cooperation*: In this case, the model consists of intracell cooperation between users by exchanging CSI information and probabilistic interference avoidance by each CCD without any exchange of information between CCDs. An example with two CCDs is shown in Fig. 12.2.

12.3 INTRACELL COOPERATION: DISTRIBUTED SCHEDULING

The cooperative intracell distributed scheduling model described in Section 9.2 is applied. Algorithm 9.1 is used in each cell for distributed resource allocation with quantized CSI. The CSI quantization method described in Section 9.2.2 is implemented.

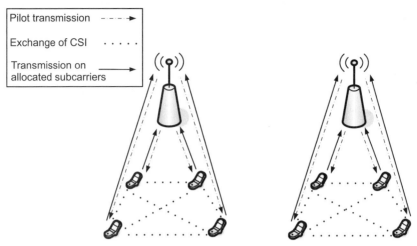

Figure 12.2   System model in the case of two noncooperating CCDs.

To implement Algorithm 9.1 in a distributed way, users are not assumed to be aware of the interference information. They are aware of the quantized CSI information exchanged with other users. Hence, when users implement Algorithm 9.1, the following expression is used in (9.6) to calculate the rate:

$$R(\mathbf{P}_{\mathbf{k}_l}, \mathcal{I}_{\text{sub},k_l}|\hat{\mathbf{H}}) = \sum_{i=1}^{N} \alpha_{k_l,i,l} \frac{B}{N} \cdot \log_2\left(1 + \beta\hat{\gamma}_{k_l,i,l}\right) \qquad (12.1)$$

where $\hat{\gamma}_{k_l,i,l}$ is the SNR over a single subcarrier, obtained in the case of quantized CSI $\hat{H}_{k_l,i,l}$. It is given by

$$\hat{\gamma}_{k_l,i,l} = \frac{P_{k_l,i,l}\hat{H}_{k_l,i,l}}{\sigma_i^2} \qquad (12.2)$$

The expression in (12.1) is used while implementing the scheduling algorithm since users only know the quantized CSI of other users. However, after subcarrier allocation, the rate achieved by each user is a function of its actual (not quantized) CSI in addition to intercell interference, and is given by (11.38) and (11.39).

12.4   INTERCELL INTERFERENCE MITIGATION/AVOIDANCE

This section describes the techniques used for the mitigation of intercell interference when cooperation between CCDs is allowed, in addition to the avoidance of intercell interference in the absence of cooperation. Each CCD computes a price reflecting the interference level it is receiving from its neighbors on each subcarrier. In the CCD collaboration scenario, the CCDs cooperate by exchanging the pricing information, that is, a CCD sends to its neighbors the computed price on each subcarrier. Two pricing-based cooperative techniques will be presented. The first method is based on transparent pricing where the pricing information is oblivious to the user. The users still transmit at their maximum power, but the approach will lead to an allocation of subcarriers that reduces the interference level in the system. The second method consists of informing the users of the price on each subcarrier so that they can perform pricing-based power control by reducing their power accordingly. In the absence of CCD collaboration, each CCD uses the computed price in a probabilistic interference avoidance approach, where a subcarrier is turned off with a probability that increases with the computed price.

12.4.1   Intercell Cooperation: Transparent Pricing Scheme

The transparent pricing method is described as follows:

- Each CCD measures the received interference power on each subcarrier.
- Each CCD l computes a price $\bar{c}_{i,l}$ on each subcarrier i based on the measured interference, with $0 \leq \bar{c}_{i,l} \leq 1$.
- The CCD communicates to its adjacent CCDs a vector $\bar{c}_l$ of prices over all its subcarriers.
- After receiving the pricing information from its surrounding CCDs, CCD l transmits on subcarrier i a pilot power given by

$$P^{\text{new}}_{\text{pilot},i,l} = \prod_{j \neq l, j=1}^{L} (1 - \bar{c}_{i,j}) P_{\text{pilot}} \tag{12.3}$$

with P_{pilot} a constant pilot power assumed to be transmitted by each CCD on each of its subcarriers. Since the pilot power is used by the users for estimating their CSI, when $P^{\text{new}}_{\text{pilot},i,l}$ is transmitted instead of P_{pilot}, each user will assume that it has a channel gain over subcarrier i given by

$$\bar{H}_{k_l,i,l} = \prod_{j \neq l, j=1}^{L} (1 - \bar{c}_{i,j}) H_{k_l,i,l} \tag{12.4}$$

- Users perform distributed scheduling as described in Section 12.3, while substituting $\bar{H}_{k_l,i,l}$ for $H_{k_l,i,l}$. Hence, when the interference on a subcarrier i increases, $\bar{H}_{k_l,i,l}$ will decrease compared to $H_{k_l,i,l}$, and user k_l will assume it has a worse channel on subcarrier i. Consequently, as the interference increases, subcarrier i will generally not be allocated to edge users, since they have a relatively lower $H_{k_l,i,l}$ on average, which reduces the interference they cause to neighboring cells. Users close to the cell center might have a relatively high $H_{k_l,i,l}$ and hence have subcarrier i allocated to one of them. This will reduce the interference since i is not allocated to an edge user. On the other hand, edge users will be allocated other subcarriers that are not causing interference to the neighboring cells and thus having reduced prices. In cases of very high interference, $\bar{H}_{k_l,i,l}$ might not be sufficient to have the subcarrier allocated to any user in the cell. In fact, if for any $j \neq l$, $\bar{c}_{i,j} = 1$, then subcarrier i will be turned off for all $l \neq j$.

Since this transparent pricing method does not involve explicit power control, then for each user to maximize its utility whether greedy or PF scheduling is used, it transmits at the maximum power. To simplify the distributed implementation of the algorithm, the maximum power of each user is subdivided equally among its allocated subcarriers. Hence,

$$P_{k_l,i,l} = P_{k_l,\max} / |\mathcal{I}_{\text{sub},k_l}| \tag{12.5}$$

This allows the users to perform distributed scheduling without exchanging any information about the transmit power of other users. The price $\bar{c}_{i,j}$ imposed by cell j on subcarrier i is selected to take a value in the interval [0 1] and is given by (11.42). The expression of the interference is given by (11.4) and (11.5).

12.4.2 Intercell Cooperation: Pricing-Based Power Control Scheme

Users are not aware of the intercell interference in the transparent pricing scheme and hence transmit at the maximum power. In this section, a power control scheme based on the pricing information exchanged between the CCDs is presented. For the greedy scheduling case, the following utility is used instead of the rate:

$$U_{k_l}\left(R_{k_l}(\mathbf{P}_{k_l}, \mathcal{I}_{\text{sub},k_l})|\hat{\mathbf{H}}\right) = \sum_{i=1}^{N} \alpha_{k_l,i,l}\left(\log_2\left(1 + \beta\hat{\gamma}_{k_l,i,l}\right) - C_{i,l}P_{k_l,i,l}\right) \tag{12.6}$$

where $C_{i,l} = S(\sum_{j \neq l, j=1}^{L} \bar{c}_{i,j})$, with $\bar{c}_{i,j}$ given by (11.42) and S a scaling factor to increase the price beyond [0 1]. Hence, (12.6) represents the rate scaled down by the power cost. However, conversely to Ref. [172], a price is imposed on the transmit power on each subcarrier, not on the total transmit power. This is in line with the approach of Ref. [53] for ad hoc networks. However, the utility is used here in greedy scheduling, conversely to Ref. [53] where scheduling is not used. With the power control scheme, the price $C_{i,l}$ of transmitting on subcarrier i in cell l is not embedded in the pilot power as in the transparent pricing scheme. Hence, users should be informed of the price of transmitting on each subcarrier. Since the prices are known at the CCD, they can be communicated to the users on the downlink via an appropriate control channel. This extra communication is not needed in the transparent pricing scheme. The power on each subcarrier when power control is used can be obtained by setting the derivative of (12.6) to zero. This yields

$$P_{k_l,i,l} = \left[\frac{1}{\ln(2)C_{i,l}} - \frac{1}{\beta\hat{H}_{k_l,i,l}/\sigma_i^2}\right]^+ \tag{12.7}$$

The drawback of the power control approach is that it requires the user utility to be "channel separable" [53]. This is easily achieved in the case of greedy maximization, since the user utility is the sum of the user rate on the individual subcarriers. However, due to the logarithm operation in PFF scheduling, the utility is not channel separable except in the special case of allocating a maximum of one subcarrier to each user. In addition, it is not practical to assume that each user is capable of keeping track of the previously achieved rate of all other users in order to implement PFTF scheduling. Hence, the power control approach will only be investigated in the case of greedy scheduling.

12.4.3   Interference Avoidance in the Absence of Intercell Cooperation: Probabilistic Transmission Scheme

The intercell interference avoidance technique is based on probabilistic transmission without CCD cooperation. Each CCD measures the received interference power on each subcarrier. Then, each CCD l computes a price $\bar{c}_{i,l}$ on each subcarrier i based on the measured interference, with $0 \leq \bar{c}_{i,l} \leq 1$, according to (11.42) and (11.5). Since $\bar{c}_{i,l}$ is selected to take a value in the interval [0 1], it can be used as a probability measure. Hence, at each scheduling interval, the CCD in cell l decides to stop using subcarrier i with probability $\bar{c}_{i,l}$, or continues using it with probability $(1 - \bar{c}_{i,l})$. Consequently, when the interference on a given subcarrier increases, its chances of being shut down also increase. The approach is similar to Section 11.6, except that it is applied in conjunction with the intracell scheduling approach described in Section 12.3.

12.5   RESULTS AND DISCUSSION

This section presents the simulation results obtained by applying the presented scheduling approach with different feedback bits per subcarrier.

12.5.1   Simulation Model

Although the model is presented in a general case, a scenario consisting of two CCDs is considered, thus representing a broadband wireless access hot spot where the presented schemes are implemented. In each cell, users are located within a radius of $R_c = 100$ m from the CCD. This radius is relatively limited in order to allow the signals transmitted by users on the cell edge to be within the reference sensitivity level of users at the diametrically opposed edge (the reference sensitivity level is discussed in Ref. [148]). Scenarios with four and eight users are studied. Users are located at fixed distances from the CCD in each cell, such that they are uniformly distributed in the interval [0 100] meters. Furthermore, a linear model is considered, as shown in Fig. 12.3. This model does not modify the conclusions obtained from using the presented approach. However, it leads to a simpler simulation model while representing a worst-case scenario: in an hexagonal cell model users are distributed throughout the cell area; this reduces the probability of edge users being close and causing high interference as in Fig. 12.3. The total bandwidth of 5 MHz is subdivided into 16 subcarriers as in Ref. [41]. A target BER of 10^{-6} is considered. The maximum user transmit power is considered to be 125 mW. The duration of one TTI is considered to be 1 ms. The channel gain over subcarrier i between user k_l in cell l and CCD j is given by (11.45), with the same parameter values used in Section 11.7.1. The quantization thresholds are selected as in Table 9.1. The threshold power P_{ref} has

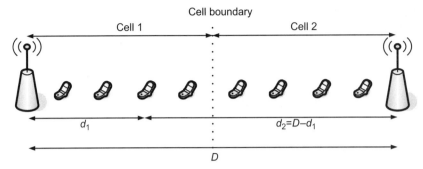

Figure 12.3 Simulation model. This figure is reprinted, with permission from IEEE, from E. Yaacoub and Z. Dawy, "A Transparent Pricing Scheme for Interference Mitigation in Uplink OFDMA with Collaborative Distributed Scheduling", IEEE International Conference on Telecommunications (ICT 2010), Doha, Qatar, April 2010 ©2010 IEEE.

the same interpretation as in Section 11.7.1. The path gain (11.45) is assumed to remain approximately constant during the duration of 10 TTIs. At each TTI, the two CCDs measure the interference, then scheduling is performed in each cell based on one of the proposed methods.

12.5.2 Greedy Allocation Results

This section presents the results of greedy scheduling with quantized feedback. Fig. 12.4 shows the rate results as a function of the distance when the transparent pricing scheme is used with $d_{\text{ref}} = 0.5 R_c$, in addition to the corresponding average subcarrier allocation for each of the users. Fig. 12.4 shows that as the number of feedback bits increases, the results become closer to the full (perfect) CSI case, with three or even two bits feedback per subcarrier achieving good results. The performance of the one bit feedback case is relatively far from the full CSI case. Despite this fact, it can be seen that it has the side effect of favoring edge users at the expense of the users nearest to the CCD. This side effect can be considered desirable in practice, since it ensures more fairness to edge users while keeping a relative advantage for the nearest users.

Tables 12.1 and 12.2 show the sum-rate results for greedy scheduling using the transparent pricing and power control schemes with four and eight users, respectively. The results of the transparent pricing scheme are denoted by TP, those of the power control scheme are denoted by PC, and the results obtained without applying any interference mitigation technique are represented by "–".

In Tables 12.1 and 12.2, both collaborative interference mitigation schemes provide enhancements over scheduling without interference mitigation. With the transparent pricing scheme, the case with $d_{\text{ref}} = 0.5 R_c$ provides more enhancements than $d_{\text{ref}} = R_c$. In fact, the case with $d_{\text{ref}} = R_c$ corresponds to a lower P_{ref}. Thus, it provides higher interference prices, which represents a higher penalty. According to (11.42) and (12.3), this leads to shutting down the most interfered subcarriers, whereas better sum-rate performance could be achieved if they were allocated to users nearer to the CCD. This statement is validated by Table 12.3, which shows the

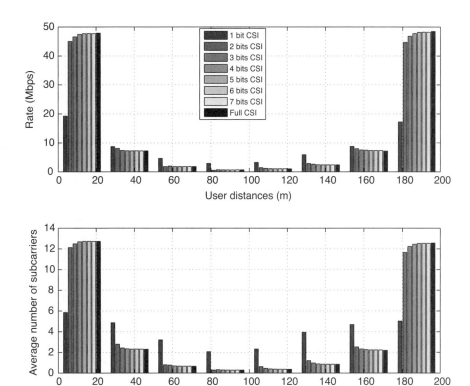

Figure 12.4   Rate and subcarrier allocation results as a function of the distance from the CCD of Cell 1—Greedy scheduling with four users per cell, $d_{\mathrm{ref}} = 0.5R_c$ and CCD collaboration using transparent pricing. This figure is adapted from E. Yaacoub and Z. Dawy, "A Transparent Pricing Scheme for Interference Mitigation in Uplink OFDMA with Collaborative Distributed Scheduling", IEEE International Conference on Telecommunications (ICT 2010), Doha, Qatar, April 2010 ©2010 IEEE.

average number of subcarriers allocated with the transparent pricing scheme. Results of the power control scheme are not shown in Table 12.3 since with that scheme, all 16 subcarriers are allocated regardless of the value of d_{ref} or the number of CSI quantization bits, similarly to the case without interference mitigation. The results

TABLE 12.1   Value of the Sum-Rate in Mbps in the Greedy Scheduling Case with CCD Collaboration: Transparent Pricing Versus Power Control in the Four Users Case

Method	–	TP	TP	PC	PC	PC	PC	PC	PC
				$S = 0.1$	$S = 0.1$	$S = 1$	$S = 1$	$S = 10$	$S = 10$
d_{ref}	–	$0.5R_c$	R_c	$0.5R_c$	R_c	$0.5R_c$	R_c	$0.5R_c$	R_c
1 bit CSI	66.22	70.56	69.80	55.36	56.94	56.08	58.43	59.54	58.28
2 bits CSI	105.27	112.30	107.93	113.73	108.59	112.61	115.3	111.30	108.24
3 bits CSI	109.13	114.74	110.95	117.54	113.03	118.18	118.36	114.73	111.85
Full CSI	110.91	116.53	112.43	118.16	114.01	119.15	119.13	115.63	112.71

TABLE 12.2   Value of the Sum-Rate in Mbps in the Greedy Scheduling Case with CCD Collaboration: Transparent Pricing Versus Power Control in the Eight Users Case

Method	–	TP	TP	PC	PC	PC	PC	PC	PC
				$S = 0.1$	$S = 0.1$	$S = 1$	$S = 1$	$S = 10$	$S = 10$
d_{ref}	–	$0.5R_c$	R_c	$0.5R_c$	R_c	$0.5R_c$	R_c	$0.5R_c$	R_c
1 bit CSI	67.56	72.33	68.32	44.40	43.04	44.76	42.63	44.33	44.88
2 bits CSI	125.97	129.32	127.32	122.44	131.93	127.66	128.75	127.47	127.62
3 bits CSI	141.86	145.55	141.81	145.09	156.60	154.55	156.16	152.18	151.84
Full CSI	150.72	154.38	150.31	150.37	161.16	159.62	160.99	156.40	157.10

of greedy scheduling with noncollaborative probabilistic interference avoidance are comparable to the case without interference avoidance. Consequently, they will not be shown here. This is explained by the fact that probabilistic scheduling leads to an excessive shutting down of subcarriers, which reduces the rate as explained above.

From Tables 12.1 and 12.2, it can be seen that all power control results outperform the case without interference mitigation, except the case of eight users, $d_{ref} = 0.5R_c$, and $S = 0.1$, which provides a comparable performance. This is explained as follows: the case of eight users corresponds to higher interference than the four users case. In addition, $d_{ref} = 0.5R_c$ leads to reduced prices compared to $d_{ref} = R_c$. Furthermore, when these prices are scaled down by $S = 0.1$ in severe interference conditions they become too reduced to affect the scheduling process by providing enough interference mitigation.

Power control results with $S = 1$ are approximately identical for $d_{ref} = 0.5R_c$ and $d_{ref} = R_c$, provide the best results in the four users case, and perform very close to the best results in the eight users case (the best results in this case are obtained with $S = 0.1$ and $d_{ref} = R_c$). This is an interesting result since it shows that the best enhancements with power control are obtained with the same prices used in the transparent pricing approach (without scaling).

Tables 12.1 and 12.2 also show that power control outperforms the transparent pricing scheme when the number of feedback bits is greater than one. This is due to the transmit power reduction used in power control, whereas transmission occurs at the maximum power with the transparent pricing scheme. This result is expected, due to the reduced complexity of intracell distributed scheduling with the transparent pricing scheme where pricing information is not communicated to the users. The enhancements obtained with the transparent pricing approach are due to a smart reallocation of subcarriers. The importance of this method is that it does not require

TABLE 12.3   Average Number of Allocated Subcarriers per Cell per TTI in the Greedy Scheduling Case with CCD Collaboration Using Transparent Pricing

No. of users	4	4	4	8	8	8
d_{ref}	–	$0.5R_c$	R_c	–	$0.5R_c$	R_c
1 bit CSI	16	15.91	15.05	16	15.81	14.14
2 bits CSI	16	15.95	15.55	16	15.97	15.52
3 bits CSI	16	15.95	15.53	16	15.97	15.52
Full CSI	16	15.96	15.59	16	15.98	15.66

the users to resort to any additional functionality than the case without interference. Hence, it preserves the simplicity of the intracell scheduling algorithm so that it can be applied in a distributed way. The distributed implementation of the power control scheme is more complex, since at every TTI, the CCD needs to inform the users of the transmission price on each subcarrier.

The greedy scheduling case represents a scenario where users are altruistic and willing to cooperate for the benefit of the network. However, in practice, it is not justifiable for an edge user to spend some of its power in exchanging feedback information while it will be unfavored by the scheduling algorithm. Hence, PF scheduling represents more motivation for the users to cooperate, since it provides a more fair distribution of resources in the network. Consequently, with PF scheduling, users are motivated to cooperate since they will be achieving their Pareto optimal rate, as determined by the Nash bargaining solution (see Section 7.2.2).

12.5.3 Proportional Fair Allocation Results

This section presents the results of PF scheduling with quantized feedback. Fig. 12.5 shows the rate results as a function of the distance when the transparent pricing scheme is used with $d_{\mathrm{ref}} = R_c$, in addition to the corresponding average subcarrier allocation

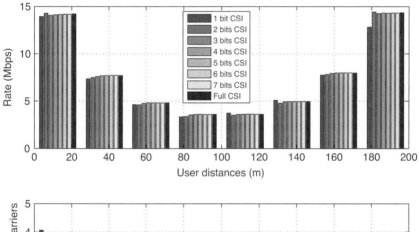

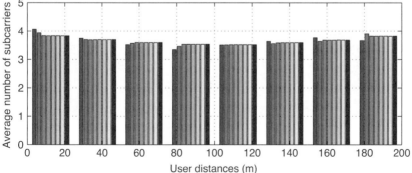

Figure 12.5 Rate and subcarrier allocation results as a function of the distance from the CCD of Cell 1—PF scheduling with four users per cell, $d_{\mathrm{ref}} = R_c$ and CCD collaboration using transparent pricing.

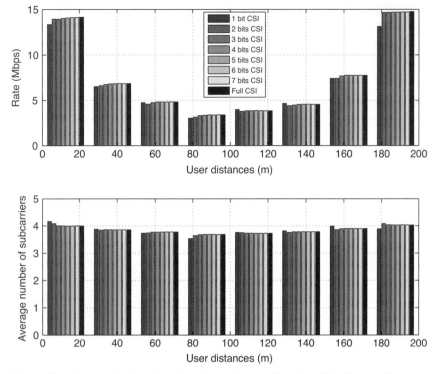

Figure 12.6 Rate and subcarrier allocation results as a function of the distance from the CCD of Cell 1—PF scheduling and probabilistic interference avoidance with four users per cell and $d_{ref} = 0.5R_c$.

for each of the users. Fig. 12.6 shows the rate results as a function of the distance when the noncollaborative probabilistic transmission approach is used with $d_{ref} = 0.5R_c$, in addition to the corresponding average subcarrier allocation for each of the users.

Figs 12.5 and 12.6 clearly show a more fair allocation of subcarriers with PF scheduling, conversely to greedy scheduling shown in Fig 12.4. Furthermore, only one bit feedback per subcarrier seems to achieve results close to the full CSI case with PF scheduling.

Tables 12.4 and 12.5 show the comparison results of PF scheduling using the transparent pricing and probabilistic scheduling schemes with four and eight users. The results of the transparent pricing scheme are denoted by TP, those of the probabilistic scheduling scheme are denoted by PS, and the results obtained without applying any interference mitigation technique (all prices set to zero) are represented by "–". Since with PF the utility used is the logarithm of the rate, the slow increase of the log function will not reflect the enhancement obtained by the presented approach, conversely to the product of the rate, which is equivalent to the sum of logarithms when used as a utility, as discussed in Section 7.2.2. Therefore, Table 12.4 shows the product of the rate, in Mbps, of the users in both cells. The values are multiplied by 10^{-6} in order to avoid excessively large numbers. Table 12.5 shows the average number of allocated subcarriers per cell per TTI.

TABLE 12.4 Value of the Rate Product ($\times 10^{-6}$) in the PF Scheduling Case: CCD Collaboration with Transparent Pricing Versus Probabilistic Scheduling Without CCD Cooperation

Method	–	TP	TP	PS	PS	–	TP	TP	PS	PS
d_{ref}	–	$0.5R_c$	R_c	$0.5R_c$	R_c	–	$0.5R_c$	R_c	$0.5R_c$	R_c
No. of users	4	4	4	4	4	8	8	8	8	8
1 bit CSI	1.54	1.78	3.01	2.47	8.72	64.92	86.95	694	466.22	3057
2 bits CSI	1.45	1.66	3.16	2.63	9.28	43.76	70.11	1024	437.56	3953
3 bits CSI	1.72	2.03	3.62	3.13	10.43	66.25	107.25	1139	609.66	4566
4 bits CSI	1.85	2.19	3.85	3.37	10.87	81.75	130.57	1263	692.90	4833
5 bits CSI	1.90	2.24	3.88	3.44	11.04	85.57	135.41	1293	728.18	4960
6 bits CSI	1.91	2.26	3.90	3.45	11.07	87.51	139.68	1294	731.51	5022
7 bits CSI	1.91	2.26	3.90	3.46	11.07	87.90	140.80	1298	732.88	5051
Full CSI	1.92	2.27	3.92	3.47	11.10	88.81	143.73	1311	742.74	5191

In Table 12.4, both the collaborative transparent pricing scheme and the non-collaborative probabilistic scheduling scheme provide considerable enhancements over scheduling without interference mitigation. With both schemes, the case with $d_{ref} = R_c$ provides more enhancements than $d_{ref} = 0.5R_c$, conversely to the greedy scheduling scenario. This is explained by the fact that with PF scheduling, a certain amount of resources must be allocated to edge users to ensure a fair allocation. Since $d_{ref} = R_c$ corresponds to higher interference prices, it leads to shutting down the most interfered subcarriers. In this case, edge users can be served with the subcarriers subjected to less interference, which enhances the results. In the greedy scheduling case, better sum-rate performance could be achieved by leaving edge users poorly served and allocating the subcarriers to users close to the CCD instead of shutting them down.

This analysis is validated by Table 12.5, which shows that the best results with PF scheduling are achieved when the number of allocated subcarriers is the least. This

TABLE 12.5 Average Number of Allocated Subcarriers per Cell per TTI in the PF Scheduling Case: CCD Collaboration with Transparent Pricing Versus Probabilistic Scheduling Without CCD Cooperation

Method	–	TP	TP	PS	PS	–	TP	TP	PS	PS
d_{ref}	–	$0.5R_c$	R_c	$0.5R_c$	R_c	–	$0.5R_c$	R_c	$0.5R_c$	R_c
No. of users	4	4	4	4	4	8	8	8	8	8
1 bit CSI	16	15.91	14.63	15.38	12.82	16	15.80	13.69	14.99	12.01
2 bits CSI	16	15.91	14.63	15.38	12.82	16	15.80	13.72	14.99	11.99
3 bits CSI	16	15.91	14.63	15.38	12.81	16	15.80	13.72	14.99	11.98
4 bits CSI	16	15.91	14.64	15.38	12.80	16	15.80	13.72	14.99	11.96
5 bits CSI	16	15.91	14.64	15.38	12.80	16	15.80	13.73	14.98	11.95
6 bits CSI	16	15.91	14.64	15.38	12.80	16	15.80	13.73	14.98	11.95
7 bits CSI	16	15.91	14.64	15.38	12.80	16	15.80	13.73	14.98	11.95
Full CSI	16	15.91	14.64	15.38	12.80	16	15.80	13.74	14.98	11.95

indicates that the presented collaborative and non-collaborative schemes were able to enhance the total network utility by shutting down the subcarriers corresponding to relatively high interference and allocating the power on the remaining subcarriers.

An interesting and unexpected outcome of this chapter is that the results of Table 12.4 without CCD collaboration outperform those with CCD collaboration. This is explained by the fact that the transparent pricing scheme does not abruptly turn off the subcarriers subjected to high interference, but rather tries to perform a smart reallocation by modifying the pilot power. Subcarriers are turned off in extreme cases (when the interference exceeds P_{ref}). On the other hand, the interference avoidance scheme without CCD collaboration turns off subcarriers with a probability that increases with the received interference level. This result is consistent with Ref. [80] where the importance of binary (on–off) power control is stressed, although scheduling is not considered. The result is also in line with Ref. [81], where a single frequency is considered and intercell coordination is applied by deactivating cells that do not offer enough capacity. Consequently, the results of this chapter complement [80, 81] by applying probabilistic on–off power control over multiple subcarriers in conjunction with user scheduling.

12.5.4 Additional Comments

In this section, some additional comments related to the simulation results are presented.

12.5.4.1 *Distributed versus Centralized Scheduling*   In a cellular system where centralized scheduling occurs at the BSs, there is no need for quantized CSI, since scheduling is performed by the BS in each cell, not by users individually. Furthermore, transparent techniques are not needed, since they are designed to simplify the scheduling operation implemented by mobile users, whereas the BSs can use more efficient techniques. In addition, the cells are not required to have a reduced coverage area, since users are not communicating with each other. However, the presented power control approach can be applied in the case of BS cooperation in the centralized scheduling scenario, whereas the probabilistic interference avoidance approach can be applied in the absence of such collaboration. Deriving these interference mitigation/avoidance techniques for centralized multicell scheduling was investigated in Chapter 11. Hence, the results of Tables 12.1–12.5 with full CSI correspond to the centralized scheduling scenario. From these tables, it can be seen that the presented interference mitigation/avoidance schemes were able to approximate the centralized scheduling case by performing distributed collaborative intracell scheduling using a limited number of CSI quantization bits.

12.5.4.2 *Transition from One to Two Feedback Bits*   Table 12.4 shows that the results are enhanced as the number of feedback bits increases, with an exception that occurs sometimes at the transition from one bit to two bits feedback. This is explained by the side effect of the one bit feedback case as discussed in Section 12.5.2: for one or two bits, the number of quantization values is still relatively far from reflecting accurately the behavior of the full CSI case. The "side effect" of one bit

quantization that led to more fairness in greedy scheduling is also noticeable with PF scheduling when the number of bits does not exceed two. This can be seen in Figs. 12.4–12.6 where the one bit feedback case provides higher rate to edge users while considerably reducing the rate of the nearest user to the CCD. This behavior is sufficient to outperform the two bits case but not the higher quantization orders. In fact, when the number of bits increases beyond two, the results of Table 12.4 increase toward the full CSI value.

12.5.4.3 Feedback Overhead The number of feedback bits leading to the best results when PF scheduling is used does not represent a considerable overhead. In Fact, Figs. 12.5 and 12.6 show that the edge users can achieve an average rate above 3 Mbps. With 16 subcarriers and seven bits feedback per subcarrier every 1 ms, the feedback overhead is 112 kbps, which represents around 3.7% for edge users, and less for users closer to the CCD. When five bits feedback are used, the feedback overhead is 80 kbps, which represents around 2.7% for edge users. This is not the case when greedy scheduling is used, since the overhead is high for edge users, as seen in Fig. 12.4. In fact, Fig. 12.4 shows that the rate of edge users is on the order of the feedback overhead.

12.6 PRACTICAL ASPECTS

This section discusses some practical implementation aspects of the investigated techniques. Some scenarios suitable for applying the presented methods are described.

12.6.1 Application in a Local Area Network

Due to the reduced cell coverage required to implement the intracell distributed scheduling technique, it lends itself to a suitable implementation in an OFDMA-based wireless local area network (LAN).[5] The CCD in this case plays the role of an access point (AP). In the case where more than one cell is needed, a reuse factor greater than one is usually used due to the reduced cell radius in wireless LANs. This is the case, for example, in 802.11, where 3 different frequencies are used by neighboring APs [160]. Fig. 12.7 shows a scenario with a reuse factor of three where CCDs labeled with the same letter use the same set of subcarriers.

12.6.2 Application in a Distributed Base Station Scenario

In a cellular network, the CCD in the presented techniques could act as an RRH in a DBS system (e.g., Chapter 8), where it connects the users to the central BS in the cell. The central BS allocates subcarriers to the different RRHs with each subcarrier

---

[5] It should be noted that the discussed application in an OFDMA-based local area network does not have any relation to existing standards, for example, the WLAN 802.11 standard and its different variations (a/b/g).

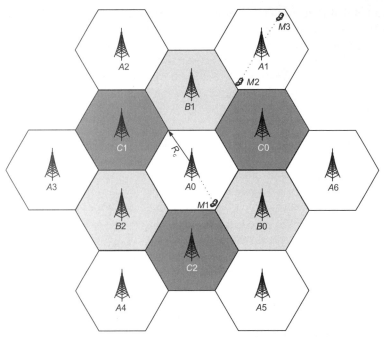

Figure 12.7 Multicell network with a reuse factor of 3.

allocated exclusively to a single RRH within a given cell. Furthermore, the central BS acts as an umbrella covering the cell sections that are not covered by the different RRHs. Consequently, it uses the subcarriers not allocated to the RRHs. Fig. 12.8 shows a model with two cells applying the DBS concept with three RRHs per cell.

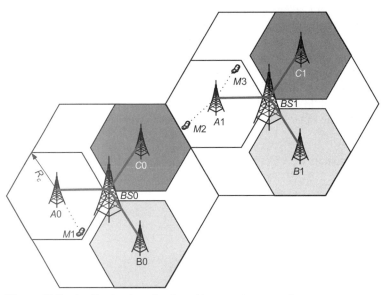

Figure 12.8 Application in a distributed base station scenario.

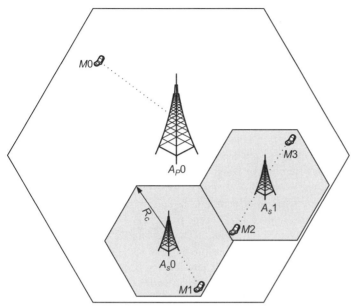

Figure 12.9 Application in a cognitive radio network with one primary cell and two secondary cells.

12.6.3 Application in a CR Network

In the presented schemes, the CCDs could be used as central controllers for scheduling secondary users in a CR network: each CCD in a CR network can sense the medium for the presence of primary users and determine the available subcarriers and then announce these subcarriers to users within its coverage range by transmitting the appropriate pilot signals. This is equivalent to turning off the subcarriers on which primary users are transmitting. Mobile users can in this case apply the same functionalities as in the presented techniques. Fig. 12.9 shows a CR scenario where A_p0 represents a BS serving primary users whereas A_s0 and A_s1 represent two CCDs in a CR network using the same subcarriers as A_p0.

12.6.4 Application in a Network with Femtocell Deployment

In the case of femtocell deployment, the techniques discussed in this chapter can be applied for interference mitigation between neighboring femtocells. The probabilistic on–off scheduling method and the transparent pricing scheme lend themselves to a straightforward implementation with femtocells. In case there is a central entity that controls femtocells, for example, a CCU inside a building that governs the resource allocation process of the femtocells inside that building, then the scenario is similar to that of an indoor DBS. In this case, a CCU coordinates the interference mitigation process during resource allocation, as discussed in Chapter 11, Section 11.4.

The discussion above does not take into account the femtocell access type. In fact, femtocell access can be either public or private. In public or open access, any user within the coverage of a femtocell can access that femtocell, even if it is associated with another femtocell (e.g., a femto BS located in its home). In private or closed access, only the user who installed the femtocell can have access rights or can grant these rights, for example, to devices and users in the apartment of the femtocell owner. Although the femtocell access method, whether open public or closed private, affects the resource allocation process and thus has an impact on network performance [175–178], the methods presented in this chapter can be applied with any access method.

However, there are additional challenges that exist in a femtocell deployment scenario. The interference can originate not only from neighboring femtocells but also from the external macrocell network. In this case, the probabilistic on–off scheduling approach can be used by a femto BS since in general no coordination exists between the indoor femto BSs and the outdoor BS covering the macrocell.

Furthermore, the performance of macrocell networks depends on the density of femtocells and hence on the interference they are causing [141, 179, 180]. In Ref. [181], it was shown that interference leakage to outside the buildings is tolerable if the building penetration loss is high enough. The work of Ref. [181] is performed in a downlink scenario where the transmissions of femtocells cause interference to macrocell users located near the buildings where femtocells are deployed. A similar downlink scenario was studied in Refs [179, 181], where power control techniques are applied in order to limit femtocell interference to macrocell users. In Ref. [145], the authors study the coverage and capacity gains from WiMAX femtocells using system-level simulations for different macrocell settings. Sensing techniques from the cognitive radio literature are adopted in order to determine the presence of macrocell users. Each femtocell senses the presence of a macrocell user. In case of its absence, the femtocell transmits at high power levels to ensure best QoS for indoor users. Although the techniques presented in Refs [145, 179, 181] are related to downlink, the scenarios studied can be extended to accommodate uplink transmissions. For example, if a femto BS senses that a macrocell user is close, the uplink transmissions of that macrocell user will interfere with the uplink transmissions of femtocell users. Hence, the femto BS can schedule the transmissions of femtocell users on other subcarriers. In the absence of macrocell users, the femto BS can indicate to the femtocell users to transmit at their maximum power.

Other techniques are applicable to both downlink and uplink. In Ref. [182], a self-organizing frequency assignment scheme is developed in which the femtocell uses sensing techniques of the air interface to dynamically allocate subcarriers to users, thereby enhancing system capacity. Additionally, fractional frequency reuse has been adopted as a resource allocation scheme for femtocells that can provide interference avoidance for coexistence with macrocells [183–185].

12.6.5 Distributed Multicell Scheduling without User Cooperation

This chapter investigated multicell scheduling in a distributed scenario with intracell user cooperation. In this section, indications are given on the implementation of the scheduling techniques described in this chapter to the case of distributed scheduling without intracell user cooperation presented in Chapter 10.

- Power control can be applied by having all the users reduce their total transmit power according to a price communicated to all users by the CCD.

- The transparent pricing scheme can be applied, since it leads to changing the priority levels of the users by altering their CSI estimation, and thus affecting the resource allocation in a way to reduce the interference.

- Probabilistic scheduling without CCD cooperation lends itself easily to distributed scheduling without intracell user cooperation, since the method described in Chapter 10 can be applied as is on the subcarriers that are still "on" after applying the probabilistic on–off scheduling approach.

12.7 SUMMARY

Distributed uplink scheduling in OFDMA systems was investigated in the case of intracell user cooperation. On the intercell level, the scenarios with the presence and absence of CCD collaboration were both studied. Cooperation between mobile users was implemented using limited feedback of channel state information, whereas cooperation between CCDs was implemented using the exchange of interference information via pricing. A transparent pricing scheme where the users are not aware of the interference information was presented in addition to a power control scheme for interference mitigation. In the absence of CCD collaboration, a probabilistic scheme was presented for interference avoidance where a subcarrier is shut down with a certain probability that increases with the interference level received on that subcarrier. Results close to the full CSI case were achieved with a limited number of feedback bits. The power control scheme led to enhanced results in the greedy maximization case. However, in a distributed scheduling scheme with user collaboration, proportional fair scheduling presents more incentives for user cooperation. When combined with proportional fair scheduling, both the transparent pricing scheme and the probabilistic interference avoidance scheme led to enhanced results in the presence of intercell interference.

CHAPTER 13

SCHEDULING IN STATE-OF-THE-ART OFDMA-BASED WIRELESS SYSTEMS

In this chapter, a scheduling overview of the state-of-the-art OFDMA-based wireless communications networks, namely WiMAX and LTE, is presented. Furthermore, the research trends concerning the next generation development of these systems are discussed. In addition, the relation of the scheduling techniques presented in this book to the state-of-the-art standards is analyzed. The chapter is organized as follows. Section 13.1 presents an overview of WiMAX scheduling. Section 13.2 presents an overview of LTE scheduling. An extension of single cell OFDMA scheduling using Algorithm 7.2 (see Section 7.3) to single carrier FDMA (SCFDMA), used in the LTE uplink, is discussed in Section 13.3. Interference mitigation in LTE uplink is investigated in Section 13.4. Finally, Section 13.5 summarizes the chapter.

13.1 WIMAX SCHEDULING OVERVIEW

WiMAX is a broadband wireless technology that supports fixed, nomadic, portable and mobile access. To meet the requirements of different types of access, two versions of WiMAX have been defined [186].

- 802.16-2004 WiMAX, based on the 802.16-2004 version of the IEEE 802.16 standard and on ETSI HiperMAN. It uses OFDM and supports fixed and nomadic access in line of sight (LOS) and nonline of sight (NLOS) environments.

- *802.16e WiMAX*: Optimized for dynamic mobile radio channels, this version is based on the 802.16e amendment and provides support for handoffs and roaming.

OFDMA is used for uplink and downlink transmission in WiMAX. WiMAX supports both FDD and TDD modes. In FDD, a fixed duration frame is used for uplink and downlink transmissions, since the transmissions occur on different frequencies.

Resource Allocation in Uplink OFDMA Wireless Systems: Optimal Solutions and Practical Implementations, Elias E. Yaacoub and Zaher Dawy.

In TDD mode, uplink and downlink transmissions take place on the same frequency. The duration of the TDD frame is fixed, but it is not necessarily divided into two equal parts: the bandwidth allocated to the downlink versus the bandwidth allocated to the uplink can change and is a system parameter. There are variable frame durations [42], but in mobile WiMAX the duration is fixed to 5 ms [187].

In WiMAX, the 192 data OFDM subcarriers are distributed in 16 subchannels of 12 subcarriers each. Each subchannel is made of four groups of three adjacent subchannels each [42]. Two types of distribution modes are available:

- *Diversity Permutations*: These are distributed permutations. The subcarriers are distributed pseudorandomly. This type includes full usage of the subchannels (FUSC) and partial usage of the subchannels (PUSC). The main advantages of distributed permutations are frequency diversity and intercell interference averaging, since they reduce the probability of using the same subcarrier in adjacent sectors or cells. However, this leads to difficulties in channel estimation since the subcarriers are distributed over the available bandwidth. For example, the steps of PUSC permutation mode are considered below:

  1. *Divide the Subcarriers into Clusters*: Subcarriers are subdivided into clusters of adjacent subcarriers.

  2. *Renumber the Clusters*: The clusters are given logical numbers instead of their physical numbers.

  3. *Gather Clusters into Six Major Groups*: The renumbered clusters are gathered in six major groups, using the logical number.

  4. *Allocate Subcarriers to Suchannels*: A subchannel is made of the data subcarriers of two clusters. Each subchannel may have subcarriers in only one major group. All the subcarriers of one subchannel belong to the same OFDMA symbol. The allocation is done according to predefined formulas [188].

- *Contiguous Permutations*: These are called adjacent permutations since they consider a group of adjacent subcarriers. With this type, channel estimation is easier since the subcarriers are adjacent, and hence allocations can be made based on the part of the bandwidth having the best channel conditions. The adaptive modulation and coding (AMC) mode is part of this permutation type.

Two WiMAX topologies are defined: point to multipoint (PMP) and mesh. In the PMP mode, traffic can take place only between the BS and the subscriber stations (SSs) associated to it (SS is the WiMAX standard term for mobile devices). In the mesh mode, SSs can act as relays and the traffic is routed via SSs in a multihop fashion until it reaches the BS. Communication can even take place only between SSs [188]. In mesh mode, an SS is identified as the Mesh BS. The Mesh BS issues a Node ID to an SS upon its request. The Node ID is the basis of SS identification in mesh mode. Mesh topology supports only TDD duplexing. In a WiMAX mesh network, scheduling is performed by the Mesh BS in a centralized way.

The network operates with a central BS in the PMP mode. It is the responsibility of the BS to perform the scheduling operations since the BS scheduler controls all the system parameters of the radio interface [188]. The BS may transmit without coordinating with other BSs. There is also a scheduler at the SS, whose role is to classify the incoming packets into the SS different connections. Five scheduling service types are defined for WiMAX in order to ensure QoS differentiation [188]:

1. *Unsolicited Grant Service (UGS)*: This service type is designed to support real-time data streams consisting of fixed-size data packets issued at periodic intervals, like the T1/E1 classical pulse coded modulation (PCM) signal, and VoIP without silence suppression.

2. *Real-Time Polling Service (rtPS)*: This service type is designed to support real-time data streams consisting of variable-sized data packets issued at periodic intervals. Hence, it is suitable for motion pictures expert group (MPEG) video transmission.

3. *Nonreal-Time Polling Service (nrtPS)*: The nrtPS is designed to support delay-tolerant data streams consisting of variable size packets having a minimum data rate requirement. This the case of file transfer protocol (FTP) transmissions.

4. *Best Effort (BE)*: This service is designed to support data streams for which no minimum service guarantees are required, for example, email.

5. *Extended Real-Time Polling Service (ertPS)*: This scheduling service type builds on the efficiency of UGS and rtPS. The ertPS allocations are dynamic, conversely to UGS allocations that are fixed in size. This scheduling type is suitable for variable rate real-time applications having data rate and delay requirements, like VoIP without silence suppression.

The first four service types were defined in Ref. [42] and the fifth was added in Ref. [187]. As for WiMAX mesh networks, they are based on TDD, and users access the channel based on TDMA [189, 190]. WiMAX mesh networks support centralized and distributed scheduling [190]: in centralized scheduling a node is selected to play the role of a BS and is denoted by mesh BS. In the distributed mode, nodes compete on TDMA time slots [190]. Current research on WiMAX mesh networks includes efficient centralized and distributed scheduling algorithms [190], in addition to interference aware scheduling [189].

In WiMAX resource allocation, the input to the scheduler at the BS is the CSI, or equivalently the channel quality indicators (CQI, which is the term most widely used in the standards). For the uplink, BSs estimate the channel directly from the pilot signals or other reference signals transmitted by the mobile devices, in both TDD and FDD. For the TDD downlink, the estimates made in the uplink could be used for downlink transmission. For the FDD downlink, channel information is obtained at the BS via feedback from the mobile terminals. In the WiMAX frequency domain, a minimal subcarrier group for AMC contains eight data subcarriers and one pilot subcarrier [191].

The output from the scheduler consists of subchannel allocation in addition to the power levels and MCS used on a subcarrier basis. The subcarriers within a subchannel follow the same power level and MCS. However, in WiMAX, different time–frequency units (called bins) can be used. Thus, a bin can be adjusted to correspond to a single subcarrier over a certain number of transmission time intervals (TTIs) [191] and hence power levels and MCS selection can be done on a subcarrier basis.

To determine the power level and MCS on the subcarriers from the reported CQI, a known method is the exponential effective SIR mapping (EESM). EESM is used to estimate the performance of the demodulator in a channel with frequency selective signal and/or noise. Hence, the EESM is a channel-dependent function that maps power level and MCS level to SINR values. This allows using this mapping in order to predict the effect of MCS and power level modification. Thus, mobile terminals report the effective SINR to the BS, and the BS decides what modulation and coding to use and with what power level.

In general, resource allocation algorithms are not imposed by the standards. However, RR, max *C/I* (greedy), and PF scheduling are supported by WiMAX, with PFTF being the most widely implemented [191, 192].

13.1.1 Enhancements in the Next Generation of WiMAX

WiMAX Release 1.0 includes only TDD operation. Both TDD and FDD are supported in Release 1.5 [193]. The next generation of WiMAX, corresponding to Release 2.0 is based on IEEE 802.16m, completed in 2010 and targeted for certification/deployment in 2011/2012. The main enhancements expected in 802.16m are as follows [193]:

- More advanced MIMO solutions, including higher order MIMO and multiuser MIMO.

- Higher peak user data rates due to the use of wideband carriers (including 20 MHz) and multicarrier aggregation.

- Enhanced coverage in environments with high interference with improved control channel.

- Faster MAC signaling leading to reduced latency.

- Support for higher mobility through a faster feedback mechanism and link adaptation.

- Flexible spectrum deployments: support of both TDD and FDD in contiguous and noncontiguous frequency bands.

- Improved interworking and coexistence with other networks such as 3G, WIFI, and Bluetooth.

- Support for multihop relay and femtocells.

- Enhanced power saving techniques.

The enhancements of WiMAX Release 2.0 are required to be backward compatible with previous releases [193].

The superframe is a new concept introduced in 802.16m [194]. A superframe has 20 ms duration, and comprises four radio frames of 5 ms each. A frame is composed of eight subframes of 0.617 ms each. The use of subframes reduces the air-link access latency from 18.5 ms in the reference system to less than 5 ms in Release 2.0. One TTI corresponds to a multiple of a subframe duration, with the default being one subframe, that is, a TTI duration is 0.617 ms [194]. The modulation schemes used in

802.16m are QPSK, 16QAM, and 64QAM, similarly to 802.16e. However, there is more granularity in the MCS schemes.

13.1.2   Intercell Interference Issues in WiMAX

WiMAX allows reuse 1 deployments, but most of the deployed systems apply reuse 3. In addition, fractional frequency reuse (FFR) is supported [192, 195, 196]. To avoid performance degradation at the cell edge with reuse 1, interference mitigation techniques are needed.

Intercell interference coordination (ICIC) in WiMAX is possible with the BSs exchanging signaling information over the R8 interface [196, 197]. In the downlink, WiMAX ICIC can be used to adjust subchannel allocation and transmit power based on the interference pattern received from neighboring cells, which leads to an increase in cell edge user rate [196]. In the uplink, ICIC can be used to adjust the transmit power of mobile devices based on load information exchanged between the BSs. This load information can be obtained by exchange of scheduling information and measurement reports from mobile devices [196].

13.1.3   Relation of the Work in this Book to WiMAX Scheduling

In this section, the applicability of the various scheduling concepts presented in this book to the WiMAX standard is discussed.

- The optimal theoretical solutions derived in Chapters 3 and 4 represent a reference to which practical solutions can be compared. Their practical implementation in WiMAX can take place if the BSs can compute the Lagrangian parameters during an initial training period then use them as long as the channel pdf does not change. It was shown in Chapter 3 that this computation can be done online and that convergence can occur after a limited number of iterations. For example, Fig. 3.5 shows that near optimal results can be achieved after 10 TTIs (10 ms with 1 ms TTIs). This corresponds to 50 ms in 802.16e (5 ms TTIs) and 6.17 ms in 802.16m (0.617 ms TTIs). The BSs do not need to keep track of the channel pdf, but they can recompute the Lagrangian parameters periodically. Performance will not be affected considerably unless the recomputation period is too large whereas the mobiles are moving too fast. In fact, for fixed and nomadic devices, a long time can elapse before recomputation is needed.

- The suboptimal algorithms presented in Chapters 6 and 7 are designed for OFDMA. Hence, they are applicable to WiMAX scheduling. However, they are mainly applicable to the BE service and to the nrtPS. They can be applied to other services by appropriately defining additional constraints, for example, delay. The subdivision of subcarriers into resource blocks (RBs) and resource allocation on an RB basis corresponds to the AMC mode of WiMAX. It should be noted that the term "RB" is borrowed from the LTE standard, and its equivalent WiMAX name is "subchannel." The number of RBs, the number of subcarriers per RB, and the TTI duration are parameters that can be tuned in the algorithm to reflect the values used in the WiMAX standard: the 192 subcarriers can be subdivided into 16 RBs (subchannels) of 12 subcarriers each, with the duration of one TTI being 5 ms in 802.16e and 0.617 ms in 802.16m.

- The application of DBS scheduling presented in Chapter 8 to WiMAX is straightforward: several RRHs deployed throughout the cell are required to be connected to a central BS. Scheduling can then take place without the requirement of standard modifications.

- The distributed scheduling approach with user cooperation presented in Chapter 9 could represent an enhancement to the WiMAX mesh network mode with centralized scheduling. The CCD of Chapter 9 would play the role of a mesh BS in this case. The noncollaborative approach of Chapter 10 could correspond to a modification of distributed scheduling in WiMAX mesh networks. The modification consists of having the users compete on OFDMA subcarriers rather than TDMA time slots.

- For multicell scheduling, ICIC in WiMAX is possible with the BSs exchanging signaling information over the R8 interface [196, 197]. The techniques presented in Sections 11.5 and 11.6 could be implemented in order to reduce the interference caused to other cells.

13.2   LTE SCHEDULING OVERVIEW

The UMTS long-term evolution (LTE) standard will stretch the performance of 3G technology, in order to meet user expectations in a 10-year perspective and beyond [198]. Goals for the evolved system include support for improved system capacity and coverage, high peak data rates (100 Mbps downlink and 50 Mbps uplink), low latency (10 ms round-trip delay), reduced operating costs, multiantenna support, flexible bandwidth operations, and seamless integration with existing systems. To meet these requirements, LTE downlink is based on OFDMA due to its immunity to intersymbol interference and frequency selective fading [198]. However, for the LTE uplink, single carrier frequency division multiple access (SCFDMA), a modified form of OFDMA, is used. Although it has similar throughput performance and essentially the same overall complexity as OFDMA, its principal advantage is its lower peak-to-average power ratio (PAPR) [199].

As in OFDMA, the transmitters in SCFDMA use different orthogonal frequencies (subcarriers) to transmit information symbols. However, in order to reduce the PAPR problem, they transmit the subcarriers sequentially, rather than in parallel [199], as shown in Fig. 13.1. High PAPR is problematic for uplink transmission where the mobile transmission power is usually limited. Due to the high PAPR value, the mean transmit power will be limited by the linear range of power amplifiers. The required backoff reduces the average transmitted power significantly and hence reduces the link budget in the uplink. This is very crucial for cell edge users due to their large path loss. For PAPR reduction, 3GPP-LTE agreed on using SCFDMA transmission with cyclic prefix in the uplink where frequency domain generation of the signal by a DFT precoding followed by an IFFT structure is assumed [98], as shown in Fig. 13.2 [199].

OFDMA versus SCFDMA

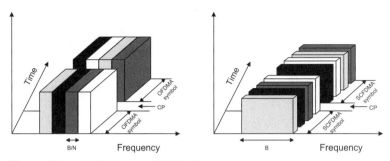

Figure 13.1   OFDMA versus SCFDMA.

Relative to OFDMA, SCFDMA reduces considerably the envelope fluctuations in the transmitted waveform. Therefore, SCFDMA signals have inherently lower PAPR than OFDMA signals [199]. However, in cellular systems with severe multipath propagation, the SCFDMA signals arrive at a base station with substantial intersymbol interference. The base station employs adaptive frequency domain

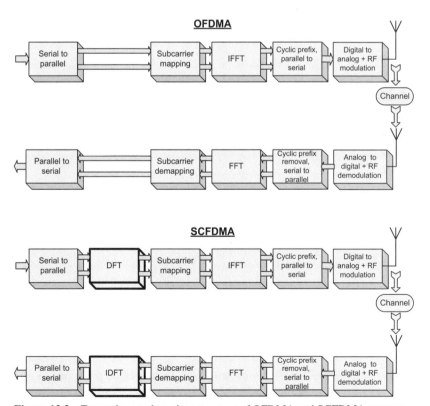

Figure 13.2   Transmitter and receiver structure of OFDMA and SCFDMA systems.

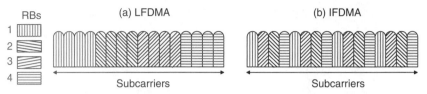

Figure 13.3 LFDMA versus IFDMA.

equalization to cancel this interference [199]. This arrangement reduces the burden of linear amplification in portable terminals at the cost of complex signal processing (frequency domain equalization) at the base station. SCFDMA has two types of subcarrier mapping [200]: Localized FDMA (LFDMA) and interleaved FDMA (IFDMA). In LFDMA, the scheduler assigns consecutive subcarriers to convey information from a particular user. In IFDMA, users are assigned subcarriers that are distributed over the entire frequency band in order to avoid allocating adjacent subcarriers that are simultaneously in a deep fade. Both mappings are shown in Fig. 13.3. IFDMA was not included into the LTE standard due to slight performance disadvantages caused by the requirements of channel estimation accuracy [41]. Hence, in the remainder of this chapter, when referring to SCFDMA, only LFDMA will be considered. In LTE, the available spectrum is divided into resource blocks consisting of 12 adjacent subcarriers. The duration of one transmission time interval is 1 ms, consisting of two 0.5 ms slots called subframes [41, 116]. In each slot, six or seven symbols are transmitted over an RB, depending whether an extended or a normal cyclic prefix (CP) is used. An extended CP corresponds to six symbols whereas the normal CP corresponds to seven symbols [199]. An uplink slot is shown in Fig. 13.4, where $N_{\mathrm{RB}}^{\mathrm{UL}}$ is the number of allocated uplink RBs, $N_{\mathrm{sc}}^{\mathrm{RB}}$ is the number of subcarriers per RB, and $N_{\mathrm{symb}}^{\mathrm{UL}}$ is the number of symbols per RB per slot (six or seven) [116]. Although in SCFDMA any set of RBs can be allocated to a single user, the LTE standard imposes the constraint that the RBs allocated to a single user should be consecutive [41, 83, 199]. The RB characteristics for the various LTE bandwidth options are shown in Table 13.1 [199]. The size of the IDFT for each channel bandwidth is larger than the number of occupied subcarriers due to the presence of guard subcarriers.

To facilitate the DFT implementation, the number of subcarriers allocated to a single user is subject to the following constraint [199]:

$$
\begin{aligned}
M_{\mathrm{sc}}^{\mathrm{PUSCH}} &= N_{\mathrm{sc}}^{\mathrm{RB}} \times 2^{\alpha_2} \times 3^{\alpha_3} \times 5^{\alpha_5} \leq N_{\mathrm{sc}}^{\mathrm{RB}} \times N_{\mathrm{RB}}^{\mathrm{UL}} \\
&= 12 \times 2^{\alpha_2} \times 3^{\alpha_3} \times 5^{\alpha_5} \leq 12 \times N_{\mathrm{RB}}^{\mathrm{UL}}
\end{aligned}
\tag{13.1}
$$

where $M_{\mathrm{sc}}^{\mathrm{PUSCH}}$ is the number of allocated uplink subcarriers, and α_2, α_3, and α_5 are nonnegative integers. Hence, (13.1) states that the number of RBs (up to 100 RBs) allocated to a mobile is in the set of integers that are multiples of 2, 3, and 5. This covers a broad range of possibilities. For example, in the case of 25 RBs (5 MHz bandwidth), the number of RBs is in the set: {1, 2, 3, 4, 5, 6, 8, 9, 10, 12, 15, 16, 18, 20, 24, 25}. Adaptive modulation and coding is used in LTE scheduling, with the modulation and coding schemes given in Ref. [84] and shown in Fig. 13.5.

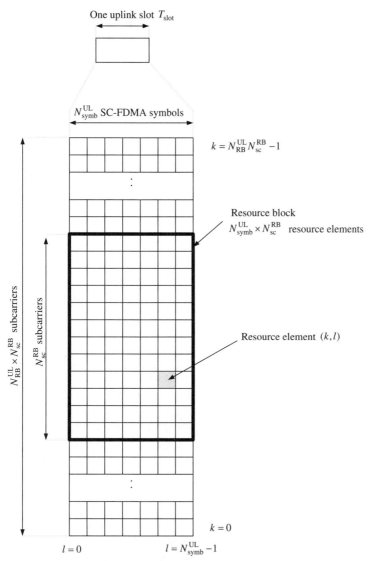

Figure 13.4 Uplink resource grid. This figure is reprinted from 3GPP TS 36.211©2008. 3GPP™ TSs and TRs are the property of ARIB, ATIS, CCSA, ETSI, TTA, and TTC who jointly own the copyright in them. They are subject to further modifications and are therefore provided to you "as is" for information purposes only. Further use is strictly prohibited.

TABLE 13.1   Resource Block Characteristics

Channel bandwidth (MHz)	1.4	3	5	10	15	20
Number of resource blocks	6	15	25	50	75	100
Number of occupied subcarriers	72	180	300	600	900	1200
IDFT/DFT size	128	256	512	1024	1536	2048

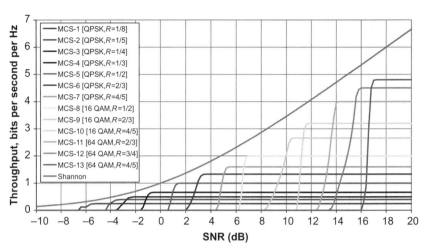

Figure 13.5 Throughput of a set of coding and modulation combinations. This figure is reprinted from 3GPP TR 36.942©2008. 3GPP™ TSs and TRs are the property of ARIB, ATIS, CCSA, ETSI, TTA, and TTC who jointly own the copyright in them. They are subject to further modifications and are therefore provided to you "as is" for information purposes only. Further use is strictly prohibited.

In LTE resource allocation, the input to the scheduler at the BS is the CQI. For the uplink, base stations estimate the channel directly from the sounding reference signals. The sounding reference signal is transmitted at the last symbol of the subframe [116]. For the downlink, channel information is acquired through explicit feedback from the mobile devices. The downlink reference signal is transmitted on the first OFDM symbol of every 0.5 ms subframe [116]. A second reference signal can be transmitted, but one is usually sufficient for FDD at low to moderate mobility, since adjacent subframes can often be used to improve channel estimation performance [201].

Reference signals are transmitted on selected subcarriers, and low complexity channel estimation (interpolation) is used to determine the CQI on the other subcarriers. Low complexity techniques include minimum mean squared error–finite impulse response (MMSE–FIR) and IFFT-based channel estimators [201].

It should be noted that MIMO and smart antenna technologies can be easily supported with OFDM, since with OFDMA each subcarrier becomes flat faded and the antenna weights can be optimized on a per subcarrier or per RB basis [201]. In the case of multiple antennas, reference signals on different subcarriers are transmitted on the different antennas. In the case of two antennas, the frequency positions of the reference signals sequence of the second antenna correspond to shifted positions of the sequence of the first antenna [116, 201]. In the case of 4×2 MIMO, the reference signals sequence corresponding to the four BS antennas is adjusted accordingly. Details can be found in Ref. [116]. Hence, to provide orthogonal reference signals for multi-antenna implementation, frequency division multiplexing (FDM) is used for different transmit antennas of the same cell. To provide orthogonality through different cells, code division multiplexing (CDM) is used for different cells.

The output from the scheduler consists of RB allocation in addition to the power levels and MCS used. The best MCS scheme in LTE can also be selected by using EESM, which is the interface between link level performance and system level simulations [192, 202]. EESM is used to derive throughput (usually determined via link level simulations) from SINR calculated on system level [192].

In general, resource allocation algorithms are not imposed by the standards. However, RR, max *C/I*, and PF scheduling are supported by LTE, with PFTF being the most widely implemented [192].

13.2.1   Enhancements in the Next Generation of LTE

After the completion of the standardization of the first release of LTE, activities on further evolution of LTE are taking place within 3GPP. Work on "LTE-Advanced" has started in April 2008. The main technology components that are currently being discussed within the framework of LTE-Advanced include [203]:

- *Carrier Aggregation*: multiple component carriers of 20 MHz are aggregated to support an overall transmission bandwidth of up to 100 MHz, thus allowing to provide very high data rates.
- *Relaying*: Relaying allows to increase coverage and reduce deployment costs by using relays instead of additional BSs when possible.
- *Extended Multi Antenna Transmission*: Increasing the data rates by allowing up to eight downlink transmission layers and up to four uplink transmission layers.
- *Coordinated Multipoint (CoMP) Transmission/Reception*: Transmission/reception is performed jointly across multiple cell sites in order to improve the cell edge performance.

The performance targets for LTE-Advanced include [204]:

- Average spectrum efficiencies of up to 3.7 bps/Hz/cell in downlink (with 4 × 4 MIMO) and 2.0 bps/Hz/cell in uplink (with 2 × 4 MIMO).
- Cell edge spectrum efficiencies of 0.12 bps/Hz/cell in downlink (with 4 × 4 MIMO) and 0.07 bps/Hz/cell in uplink (with 2 × 4 MIMO).
- Peak data rates of up to 1 Gbps in the downlink and 500 Mbps in the uplink.
- Peak spectrum efficiencies of 30 bps/Hz in the downlink (with up to 8 × 8 MIMO) and 15 bps/Hz in the uplink (with up to 4 × 4 MIMO).
- Low cost of infrastructure deployments and terminals.
- Power efficiency in the network and terminals.

13.2.2   Intercell Interference Issues in LTE

LTE allows communication between BSs over the X2 interface [85]. Intercell interference coordination between LTE BSs can be performed using the overload indicator (OI) and the high interference indicator (HII). In the uplink, the OI indicates the interference level received by the sending cell on each RB, whereas the HII indicates the occurrence (or not) of high interference on each RB [85]. The receiving BS would then try to perform scheduling while avoiding allocation on the RBs subjected to high interference in its neighbor cells. In the current LTE standard, the minimum latency for the exchange of information between BSs is 20 ms, whereas RBs are allocated on a 1 ms basis (duration of one TTI). This makes real-time processing of interference cancellation data from adjacent BSs unfeasible [86]. Other interference coordination and cancellation techniques applicable to the current LTE standard are surveyed in Ref. [86] (applicable to both uplink and downlink, unless otherwise specified): power control, static and adaptive fractional frequency reuse, MIMO (LTE downlink), multiuser MIMO (LTE uplink), space division multiple access (SDMA), interference cancellation (LTE uplink), opportunistic spectrum access, organized beamforming, sphere decoding (LTE uplink), and dirty paper decoding (LTE uplink).

Major research is ongoing for network MIMO within the framework of LTE-Advanced, such as [89]:

- Synchronization of jointly processed terminals in time and frequency, and detection under synchronization offsets.

- Multisector channel estimation, feedback of CSI to BSs, and impact of imperfect CSI on network MIMO.

- Performance of network MIMO under a limited backhaul infrastructure between cooperating BSs.

- Cooperative scheduling for network MIMO.

13.2.3   Relation of the Work in this Book to LTE Scheduling

In this section, the applicability of the various scheduling concepts presented in this book to the LTE standard is discussed.

- The optimal theoretical solutions derived in Chapters 3 and 4 represent a reference to which practical solutions, including those based on LTE, can be compared. Their implementation in LTE is limited since the optimal solution is derived for OFDMA, not SCFDMA. Although the extension to SCFDMA is straightforward when the contiguous RB constraint is not imposed, deriving the optimal solution when this constraint is enforced is not straightforward. In this case, the set of RBs should be partitioned into subsets of contiguous RBs, and the optimal mapping of subsets to users should be performed jointly with optimal set partitioning.

- The suboptimal algorithms presented in Chapters 6 and 7 are designed for OFDMA, but are also applicable to SCFDMA. In most cases, the simulation setup is consistent with LTE parameters: bandwidth, number of consecutive subcarriers in an RB, TTI duration, and so on. The algorithms can accommodate a varying number of subcarriers per RB, with the case of one subcarrier per RB corresponding to classical OFDMA scheduling. The algorithms are utility maximizing algorithms. Fairness can be imposed by an appropriate choice of the utility: for example, a logarithmic utility ensures proportional fairness. However, the algorithms do not respect the contiguous RB allocation constraint, imposed for SCFDMA in LTE in order to keep its advantage of low PAPR. A simple extension of Algorithm 7.2 to enforce this constraint is presented in Section 13.3 and the results of both algorithms are analyzed and compared.

- The application of DBS scheduling presented in Chapter 8 to LTE is straightforward, on the condition that the algorithms are adjusted to enforce the contiguous RB constraint.

- The distributed scheduling techniques, whether including user cooperation as described in Chapter 9 or without user cooperation as in Chapter 10, do not have currently a direct application to LTE. However, they represent interesting topics for investigating future coexistence between LTE and CR networks where a user uses LTE in its home network and implements the presented techniques in CR networks.

- For multicell scheduling, cooperation between BSs occurs every 20 ms in the current LTE standard over the X2 interface [85], whereas scheduling occurs every 1 ms. In LTE-Advanced, plans are included to reduce this delay and allow scheduling coordination between BSs. Hence, BS cooperation is interesting in the framework of LTE-Advanced, whereas noncooperative solutions are interesting in the framework of the current LTE standard. The iterative pricing game in the presence of a CCU presented in Section 11.4 can be implemented in practice in the case of a DBS scenario. In fact, if full reuse is allowed in the coverage area of every RRH (conversely to Chapter 8), the central BS can be aware of the CSI between all users and all RRHs, which allows an offline application of the controlled pricing game, before the scheduling decisions are communicated to the RRHs. However, for interference mitigation between cells (not just between the coverage areas of RRHs), other techniques should be derived, for example, those presented in Sections 11.5 and 11.6. The pricing-based power control approach without BS collaboration, presented in Section 11.5, is extended to the LTE framework and compared to the LTE power control scheme in Section 13.4. The purpose of power control in the uplink is to judiciously regulate the transmit power of the mobile users in order to reduce the interference caused to other cells. Joint power control and resource allocation leads to enhanced performance, as in the presented approach of Section 13.4.

13.3   SCFDMA VERSUS OFDMA SCHEDULING

In this section,[6] an extension of single cell OFDMA scheduling to SCFDMA is discussed.

SCFDMA is combined with frequency dependent scheduling in Ref. [205] and high spectral efficiency is achieved in moderate and high SNR conditions. It is also

---

[6] This section is adapted, with permission from IEEE, from E. Yaacoub and Z. Dawy, "A Comparison of Uplink Scheduling in OFDMA and SCFDMA", IEEE International Conference on Telecommunications (ICT 2010), Doha, Qatar, April 2010 ©2010 IEEE.

shown in Ref. [205] that for cell edge users with relatively low SNR, SCFDMA increases the cell edge rate. A search-tree based channel aware packet scheduling algorithm is proposed in Ref. [206] and its performance is evaluated in terms of rate and noise rise distributions. In Ref. [207], an SCFDMA resource allocation algorithm is derived based on a pure binary-integer program called the set partitioning problem. In addition, a suboptimal greedy heuristic that performs close to the proposed algorithm with lower complexity is presented. A rate comparison between OFDMA and SCFDMA is presented in Ref. [208], but it is limited to a maximum of 20 users with proportional fair scheduling. In this section, the uplink scheduling performance of both OFDMA and SCFDMA is compared in the case of continuous and noncontinuous RB allocation with different utility functions and different fading conditions for up to 64 users in a single cell scenario.

13.3.1   SCFDMA Rate Calculations

For the SCFDMA rate calculations, the Shannon upper bound is considered:

$$R_k(P_k, \mathcal{I}_{\text{sub},k}) = \frac{B|\mathcal{I}_{\text{sub},k}|}{N_{\text{sub}}} \cdot \log_2\left(1 + \gamma_k(P_k, \mathcal{I}_{\text{sub},k})\right) \tag{13.2}$$

where $\gamma_k(P_k, \mathcal{I}_{\text{sub},k})$ is the SNR of user k after MMSE frequency domain equalization at the receiver [199]:

$$\gamma_k(P_k, \mathcal{I}_{\text{sub},k}) = \left(\frac{1}{\dfrac{1}{|\mathcal{I}_{\text{sub},k}|} \displaystyle\sum_{i\in\mathcal{I}_{\text{sub},k}} \dfrac{\gamma_{k,i}}{\gamma_{k,i}+1}} - 1 \right)^{-1} \tag{13.3}$$

The rate calculations for OFDMA are performed according to (7.12), with the SNR given by (7.13). In both SCFDMA and OFDMA scheduling, each user is assumed to transmit at the maximum power, and the power is assumed to be subdivided equally among all the subcarriers allocated to that user, according to (7.14).

13.3.2   Scheduling Algorithm with Contiguous RBs

In this section, Algorithm 13.1 is presented, which is a direct extension of Algorithm 7.2 to the case of the continuous RB constraint. It enforces the constraint that the RBs allocated to a given user are consecutive.

Algorithm 13.1

> **Scheduling Algorithm with Contiguous RBs**
> **Subcarrier Allocation:**
> **Consider the set of available users $\mathcal{I}_{\text{avail\_users}} \subseteq \{1, 2, ..., K\}$. At the start of the algorithm $\mathcal{I}_{\text{avail\_users}} = \{1, 2, ..., K\}$.**
> **for k such that $1 \leq k \leq K$**

$\mathcal{I}_{\text{RB},k} \leftarrow \emptyset$
end for
for i **such that** $1 \leq i \leq N_{\text{RB}}$
    for all k **such that** $k \in \mathcal{I}_{\text{avail\_users}}$
        Compute $U_k(P_k, \mathcal{I}_{\text{RB},k} \cup \{i\})$
        if $\mathcal{I}_{\text{RB},k} \neq \emptyset$
            Compute $U_k(P_k, \mathcal{I}_{\text{RB},k})$
        else
            $U_k(P_k, \mathcal{I}_{\text{RB},k}) \leftarrow 0$
        end if
        $\Lambda_{k,i} \leftarrow U_k(P_k, \mathcal{I}_{\text{RB},k} \cup \{i\}) - U_k(P_k, \mathcal{I}_{\text{RB},k})$
    end for
    $k^* \leftarrow \arg\max_k \Lambda_{k,i}$
    if $\Lambda_{k^*,i} > 0$
        $\mathcal{I}_{\text{RB},k^*} \leftarrow \mathcal{I}_{\text{RB},k^*} \cup \{i\}$
        if $(i > 1$ **and** $k^* \neq \arg\max_k \Lambda_{k,i-1})$
            Comment: If k^* **is the same user to which RB** $i - 1$ **was allocated, keep** k^* **in** $\mathcal{I}_{\text{avail\_users}}$. **Otherwise, delete user** k^* **from the set of available users**
            $\mathcal{I}_{\text{avail\_users}} \leftarrow \mathcal{I}_{\text{avail\_users}} \setminus \{k^*\}$
        end if
    end if
end for
Power Allocation:
Apply Algorithm   6.2 for power allocation

Algorithm 7.2 has a complexity of $\mathcal{O}(N_{\text{RB}} K)$, as shown in Section 7.3. Consequently, Algorithm 13.1 has a worst-case complexity of $\mathcal{O}(N_{\text{RB}} K)$ when all the RBs are allocated to a single user. As the number of users in $\mathcal{I}_{\text{avail\_users}}$ decreases, the complexity of Algorithm 13.1 decreases due to the reduced number of users involved in the search.

13.3.3   Results and Discussion

The simulation model of Section 7.5.1 is considered. Users are uniformly distributed in the cell. The total bandwidth considered is $B = 5$ MHz, subdivided into 25 RBs of 12 subcarriers each. The maximum mobile transmit power is considered to be 125 mW. Each mobile is assumed to transmit at the maximum power, and the power is subdivided equally among all subcarriers allocated to it. The value of β is set to $\beta = 1$ in (7.12), since the expression of the SNR gap β in (6.3) corresponds to uncoded MQAM. Hence, setting $\beta = 1$ leads to a better approximation when channel coding is used. In addition, it leads to a more fair comparison with (13.2) which implicitly assumes $\beta = 1$.

In the simulation results presented, two fading scenarios are considered: the case of independent fading on each subcarrier in an RB and the case of identical fading along all the subcarriers constituting an RB. The former case corresponds to

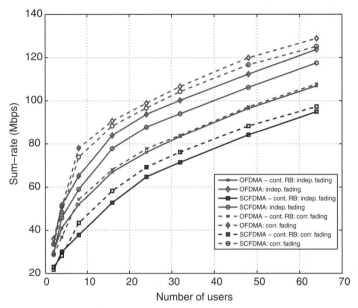

Figure 13.6 Sum-rate for the different OFDMA and SCFDMA scenarios with greedy scheduling ($U = R$).

uncorrelated fading and the latter corresponds to full correlation. Hence, the actual system performance will be between these two extremes. Algorithms 7.2 and 13.1 are applied to both OFDMA and SCFDMA in these two fading scenarios.

Fig. 13.6 shows the results for greedy scheduling ($U = R$). Considering the case of independent fading, and comparing the case of continuous RBs to the case where this constraint is not imposed, it can be concluded that around 15% degradation is obtained in the case of OFDMA and around 20% degradation is reached in the case of SCFDMA when the continuous RB constrained is imposed. This is expected since frequency diversity is invested more in the case of free scheduling without the continuous RB constraint. Comparing SCFDMA to OFDMA in the case of independent fading, it can be seen that OFDMA outperforms SCFDMA by around 11% when the continuous RB constraint is imposed and by around 5% when it is not. In OFDMA, AMC can be performed over every subcarrier, which is not the case for SCFDMA [199]. Although the Shannon capacity formula is adopted, the AMC limitation is translated by the equivalent SNR in (13.3) for the whole SCFDMA transmission, whereas for OFDMA the Shannon capacity (7.12) is applied for every subcarrier. Comparing the case of independent fading to the case of correlated fading, it can be seen that a better performance is obtained with the correlated fading case, except for the case of OFDMA with continuous RBs, where the performance is comparable. In the case of independent fading, all the subcarriers of a single RB are forced to be allocated together, although the fading might have different fluctuations on each subcarrier. However, in the case of equal fading, all subcarriers of a single RB have the same instantaneous fading value, and thus it is logical to allocate them together.

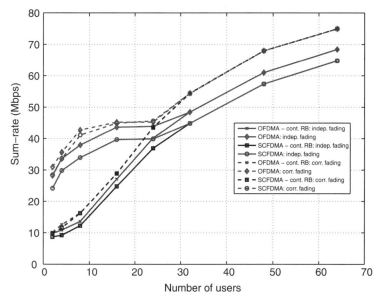

Figure 13.7 Sum-rate for the different OFDMA and SCFDMA scenarios with PFF scheduling ($U = \ln(R)$).

It should be noted that, if subcarriers were allowed to be allocated individually, better performance would be reached with independent fading since frequency diversity would be used more efficiently in this case. However, in the worst case, the difference between independent and equal (maximum correlation) fading does not exceed 10%. In practice, the actual performance would be between these two extremes.

The results for PFF scheduling ($U = \ln(R)$) are shown in Fig. 13.7. They are consistent with those of Ref. [208] for the same range of the number of users (PF scheduling with up to 20 users in Ref. [208]). The same conclusions as in the greedy scheduling case can be reached regarding the superiority of OFDMA over SCFDMA and the superiority of correlated fading over independent fading. However, it should be noted that OFDMA and SCFDMA have approximately the same performance in the correlated fading case. Furthermore, it is remarkable that the case of free scheduling without the continuous RB constraint outperforms the continuous RB case until the number of users exceeds the number of RBs, where both plots converge to the same values. This behavior is explained by the fact that the algorithms avoid having a user with zero rate, since this would make the sum of logarithms of the rates become $-\infty$, or, equivalently, set the product of the rates to zero. Hence, when the continuous RB constraint is not imposed, there is at least one RB allocated to each user. When the number of RBs is equal to the number of users, one RB is allocated per user in both algorithms. However, when the number of users exceeds the number of RBs, only the N_{RB} users (out of K users) having the best channel conditions are allocated N_{RB} RBs, one for each user. Since no user is allocated more than one RB, the continuous RB constraint does not hinder the performance of Algorithm 13.1 and both algorithms converge to the same results. It should be noted that, in the simulations, when a

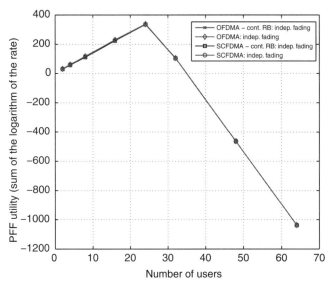

Figure 13.8 Sum-utility for the different scenarios with PFF scheduling.

user has zero rate, and to avoid dealing with $-\infty$ values, the rate of that user is set to a very low value ϵ instead, with $\epsilon = 2.2 \times 10^{-16}$ in Matlab, as in Section 7.5.5. This explains the linear decrease in the utility function $U = \ln(R)$ in Fig. 13.8 when K becomes greater than N_{RB}. In Fig. 13.8, only the case of independent fading is presented. Fig. 13.8 shows that, in terms of utility maximization, no real differences are observed between the two algorithms and between OFDMA and SCFDMA when $U = \ln(R)$.

13.4 COMPARISON TO THE LTE POWER CONTROL SCHEME

In this section,[7] interference mitigation in LTE uplink is investigated. Pricing-based power control is extended to LTE and compared to standard LTE power control.

In the SCFDMA-based LTE uplink, additional constraints should be taken into account in the scheduling process. The LTE standard imposes the constraint that the RBs allocated to a single user should be consecutive with equal power allocation over the RBs [41, 83, 199]. In addition, LTE allows intercell interference coordination between LTE BSs by communicating the OI and HII over the X2 interface [85].

The LTE power control scheme is detailed in Ref. [83], and its evaluation via simulations is described in Ref. [84]. The transmit power (in dBm) according to the

---

[7] This section is adapted, with permission from IEEE, from E. Yaacoub and Z. Dawy, "Joint Uplink Scheduling and Interference Mitigation in Multicell LTE Networks", ICC 2011, Kyoto, Japan, June 2011 ©2011 IEEE.

LTE power control scheme is given by [83]

$$P_k = \min \left[P_{k,\max}, 10 \log_{10}(M_{RB}^{PUSCH} + P_0 + \alpha PL + \Delta_{TF} + f) \right] \quad (13.4)$$

where

- M_{RB}^{PUSCH} is the number of assigned resource blocks as indicated in the uplink scheduling grant.

- P_0 is a parameter composed of the sum of an 8-bit cell specific nominal component signaled from higher layers with 1 dB resolution and a 4-bit user equipment (UE) specific component configured by the RRC with 1 dB resolution.

- $\alpha \in \{0, 0.4, 0.5, 0.6, 0.7, 0.8, 0.9, 1\}$ is a 3-bit cell specific parameter provided by higher layers.

- PL is the downlink pathloss estimate calculated in the UE.

- Δ_{TF} is signaled by the RRC and depends on the used MCS.

- f is a function signaled by higher layers

The expression in (13.4) depends on many system parameters. To assess the performance of LTE power control via simulations, the approach described in Ref. [84] is applied. The total transmit power of user k is given by [84]

$$P_k = P_{k,\max} \times \min \left[1, \left(\frac{PL}{PL_{x-ile}} \right)^{\gamma} \right] \quad (13.5)$$

where PL_{x-ile} is the x-percentile path loss (plus shadowing) value, and γ is a balancing factor for UEs with bad channel and UEs with good channel. The values of these parameters are set to $\gamma = 1$ and $PL_{x-ile} = 115$ dBm, as determined in Ref. [84] for $B = 5$ MHz.

In this section, scheduling and interference mitigation in LTE uplink are investigated. Low complexity cooperative and noncooperative schemes that are consistent with the LTE standard are presented and shown to lead to enhanced performance. The presented schemes perform joint power control and scheduling in the uplink direction. Algorithm 13.1 that enforces the LTE contiguous RB constraint while allowing several RBs to be allocated per user is used. However, Algorithm 13.1 is implemented in each cell in conjunction with the noncooperative probabilistic interference avoidance scheme and a cooperative pricing-based power control scheme based on the OI indicator. The PFTF utility $U_{k_l} \left(R_{k_l}(\mathbf{P_{k_l}}, \mathcal{I}_{RB,k_l}) \right) = \dfrac{R_{k_l}(\mathbf{P_{k_l}}, \mathcal{I}_{RB,k_l})}{D_{k_l,\text{tot}}}$, defined in Chapter 7, Section 7.4, is used.

13.4.1   LTE Multicell Interference Mitigation Schemes

As a noncooperative scheme, the probabilistic interference avoidance scheme of Section 11.6 is used. Each BS then implements Algorithm 13.1 on the subcarriers that are still on. Hence, this scheme does not require coordination between BSs since measurements are made at each BS separately without any coordination over the X2 interface.

The cooperative pricing-based power control scheme is a modification of the scheme of Section 11.5, in order to accommodate the LTE equal power constraint. Hence, in the pricing-based power control scheme, $\bar{c}_{i,j}$ computed in (11.42) is not used by cell j as in the probabilistic scheduling scheme. The price $\bar{c}_{i,j}$ corresponds to the total interference received by BS j from all its neighbors. However, it is practically difficult for BS j to determine the contribution of each of its neighboring BSs (e.g., BS l) in the total received intercell interference. Hence, a logical assumption would be to penalize all neighboring BSs equally with the average interference price. Consequently, BS j transmits a price $c_{i,j}^{(\text{avg})} = \bar{c}_{i,j}/N_{\text{Neighbors}}$ to all BSs in $\mathcal{N}_j$, the set of BSs that are neighbors to BS j, with $N_{\text{Neighbors}}$ the number of neighbors of each cell. Hence, $c_{i,j}^{(\text{avg})}$ represents the price imposed by cell j on the transmission over subcarrier i in neighboring cells. Neighboring BSs exchange the prices $c_{i,j}^{(\text{avg})}$ (e.g., on the OI indicator of the X2 interface) and then each BS l sets the transmission price on subcarrier i as $\bar{C}_{i,l} = c_{i,j^*}^{(\text{avg})} = \max_{j \in \mathcal{N}_l} c_{i,j}^{(\text{avg})}$. Hence, selecting the price sent by j^*, the neighboring BS sending the highest interference cost on i, allows BS l to accommodate the most stringent requirement of its neighbors.

The transmit power on subcarrier i in cell l is set as follows:

$$P_{k_l,i,l} = \frac{P_{k_l,\max} \cdot \max(1 - \delta \max_i \bar{C}_{i,l}, 0)}{|\mathcal{I}_{\text{sub},k_l}|} \forall i \in \mathcal{I}_{\text{sub},k_l} \tag{13.6}$$

with δ a constant used as a scaling parameter, in order to avoid excessive power reduction. BS l then uses (13.6) to compute the power in Algorithm 13.1. The power expression in (13.6) satisfies the LTE constraint of using equal power over all subcarriers. Moreover, it meets the pricing penalties received on all subcarriers by using the highest price over the subcarriers allocated to user k_l, $\max_i \bar{C}_{i,l}$, in order to reduce the maximum transmit power. In the absence of power control, $\bar{C}_{i,l} = 0$ is satisfied for all i and (13.6) corresponds to maximum power transmission.

In practice, a single interference level per RB is exchanged over the X2 interface via the OI indicator. However, since all subcarriers of an RB are allocated together and are subjected to approximately the same fading [41, 199], the received price can be applied to all subcarriers of a given RB. It should be noted that $N_{\text{Neighbors}}$ and δ are predefined constant parameters whose values are used by all BSs.

13.4.2   Results and Discussion

The channel gain of (11.45) is considered along with the parameters of Section 11.7.1. In addition, a total bandwidth of 5 MHz subdivided into 25 RBs of 12 subcarriers each is assumed. The maximum user transmit power is considered to be 125 mW. The duration of one TTI is 1 ms. The channel gain is assumed to remain approximately constant during the duration of 10 TTIs. The value of δ is set to $\delta = 0.05$.

The results are compared to the reuse 1 scheme with and without applying LTE power control in addition to the FFR 1/3 scheme illustrated in Fig. 2.2, where the

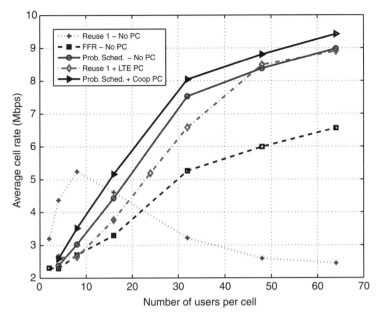

Figure 13.9 Average cell rate.

entire bandwidth is divided into four segments. Part of the RBs is used with reuse of 1 in the cell center region, whereas reuse 3 is applied in regions A, B, and C [77]. In the simulations, 13 RBs are used in the cell center, and four RBs are used in each of the regions A, B, and C (the total is 25 RBs per cell).

Fig. 13.9 shows the average cell rate results for reuse 1, FFR, and the presented probabilistic interference avoidance scheme without power control (maximum power transmission). In addition, the scenarios of reuse 1 with standard LTE power control and the probabilistic scheduling scheme with cooperative power control are also shown. In the case without power control, the probabilistic scheduling scheme has a clear superiority over the reuse 1 and FFR scenarios. It should be noted that for a reduced number of users, the reuse 1 scheme is superior since the interference level in the network is low enough to allow for full power transmission. When power control is added to the above schemes, it can be seen from Fig. 13.9 that the power control scheme used in the LTE standard (LTE PC) [83, 84] enhances the performance of the reuse 1 scheme. However, it is outperformed by the probabilistic scheduling scheme until the number of users becomes large. The best performance is obtained when the probabilistic scheduling scheme is used with the presented cooperative pricing-based power control. The superiority of this combined scheme stems from the fact that it performs interference mitigation using two distinct measures:

- A "defensive" measure characterized by the probabilistic interference avoidance scheme, where each BS reacts to the received interference level without communicating with other BSs.

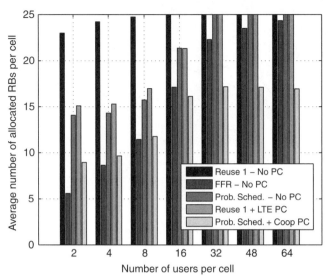

Figure 13.10 Average number of allocated RBs per cell.

- An "offensive" measure implemented via the cooperative power control scheme, where each BS penalizes its neighbors by communicating transmission prices that force the neighboring BSs to reduce their transmission power.

The scenarios of FFR with LTE power control, probabilistic scheduling with LTE power control, and reuse 1 with the presented cooperative power control were simulated and shown to have a very slight superiority over FFR without power control, probabilistic scheduling without power control, and reuse 1 with LTE power control, respectively. Hence, they are not displayed in Fig. 13.9.

Fig. 13.10 shows the average number of allocated RBs of the schemes of Fig. 13.9. For a small number of users, the FFR scheme has a low number of allocated RBs due to the limited number of available RBs in the regions of Fig. 2.2. As the number of users increases, it can be seen that the presented scheduling algorithm with probabilistic interference avoidance and cooperative power control achieves the best results in Fig. 13.9 by allocating the lowest number of RBs, as shown in Fig. 13.10. Hence, this joint approach succeeds in turning off the RBs subjected to the highest interference and judiciously concentrating the power on the other RBs.

The presented schemes are conforming to the LTE standard that provides flexibility in using ICIC mechanisms [87]. The probabilistic scheduling scheme does not require ICIC coordination between BSs. The pricing-based power control scheme requires ICIC communication via the OI indicator over the X2 interface. In LTE, although the standard does not explicitly prescribe the time scale at which ICIC should operate, ICIC is currently performed on a scale of tens to hundreds of milliseconds, whereas fast scheduling is performed each millisecond [87, 89]. Hence, in low mobility scenarios with slowly varying channels, which is a widely used assumption [87, 172, 209], the presented cooperative power control scheme is applicable to the current LTE standard. Fast channel variations, for example, every 10 ms, with ICIC performed every 1 ms is beyond the current LTE standard and may be applicable to LTE-Advanced, where faster ICIC control is being considered [89].

13.5 SUMMARY

In this chapter, an overview of resource allocation in WiMAX and LTE was presented. Insights on the most relevant research directions for the next generation of both systems were indicated. The work in this book was related to WiMAX and LTE scheduling. The following scenarios were investigated in detail:

- *Single Cell LTE Scheduling*: Single cell uplink scheduling in OFDMA and SCFDMA systems was considered. OFDMA was found to outperform SCFDMA for the same number of RBs and the same maximum transmission power. The case of independent fading over the subcarriers of a single RB was compared to the case of equal fading corresponding to full correlation, and the case of equal fading was found slightly superior. Scenarios with and without the continuous RB constraint were also compared. The constraint was shown to reduce the performance by an amount not exceeding 20% in the case of sum-rate maximization. In the case of a logarithmic utility function, both scenarios converged to the same results when the number of users exceeded the number of RBs.

- *Multicell LTE Scheduling*: Interference mitigation in uplink multicell LTE networks was investigated. The noncooperative probabilistic interference avoidance scheme of Chapter 11 was adopted. In addition, a pricing-based power control scheme was studied in the presence of BS cooperation. The interference mitigation schemes were implemented in conjunction with a low complexity scheduling algorithm. Performance results demonstrate notable gains compared to reuse 1 with standard LTE power control.

CHAPTER 14

FUTURE RESEARCH DIRECTIONS

In this chapter, insights on more advanced topics in OFDMA resource allocation are presented, and interesting directions for future investigation are discussed in Sections 14.1–14.8.

14.1 RESOURCE ALLOCATION WITH MULTIPLE SERVICE CLASSES

In this book, the problem of the varying QoS constraints required by different traffic types was tackled by including minimum rate constraints in the formulation and solution of the weighted ergodic sum-rate maximization problem treated in Chapter 3. Furthermore, the role of the weights in ensuring different priority levels to different users as indicated by higher layers was explained in Chapters 3 and 4. In addition, fair allocation of resources was treated by dedicating large sections of this book to proportional fair scheduling (e.g., in Chapters 7 and 9). However, all the users in a network were assumed to use the same service type. Investigating scheduling in a mixed network scenario represents an interesting extension of this work. Downlink resource allocation with mixed traffic consisting of constant bit rate and best effort users in OFDMA was investigated in Ref. [210], where an upper bound and a suboptimal algorithm were derived. The uplink formulation and solution are under investigated in the literature. Another interesting extension would be to incorporate delays in the problem formulation, in order to model the delay constraints of real-time services. This leads to cross-layer extensions of the presented resource allocation algorithms by combining them with network layer packet scheduling, where packet waiting time, packet delay, queue stability, and buffer overflow all come into play. Packet scheduling in OFDMA is treated in Refs [23, 206, 211, 212]. However, a cross-layer formulation involving optimal scheduling combined with queuing theory concepts represents an interesting problem worthy of further investigation.

14.2 NETWORK MIMO

Open research areas include network MIMO where a user is served by several BSs within its surroundings. Thus, network MIMO requires tight coordination of trans-

Resource Allocation in Uplink OFDMA Wireless Systems: Optimal Solutions and Practical Implementations, Elias E. Yaacoub and Zaher Dawy.
© 2012 by the Institute of Electrical and Electronics Engineers, Inc. Published 2012 by John Wiley & Sons, Inc.

mission and reception at multiple BSs [213]. It is investigated in LTE-Advanced under the name of coordinated multipoint (CoMP). Network MIMO has interesting applications in a distributed base station scenario, where the central BS is able to know the channel state information between all users and all remote radio heads. Hence, the BS can apply network MIMO by transmitting and receiving the signals of a given user using several remote radio heads.

In fact, in Chapter 8, we outlined the importance of DBSs in providing remote antennas at a close proximity to the mobile users. A more challenging scenario that builds on the results of Chapter 8 would be to compare the CBS scenario to DBS assuming the same number of receive antennas in both cases. In this case, the CBS can exploit the additional antennas to separate the users, implement MIMO, diversity techniques, and possibly use interference cancellation to improve the system performance. On the other hand, the DBS could use the distributed antennas to expand the concept of CoMP and network MIMO to the coverage area of a single BS. In this case, the joint DBS–CoMP combination would represent a scenario of efficient implementation of network MIMO, since the central BS would have full knowledge of the network state within its cell and complete control over the antennas involved in the network MIMO process.

Another open area is multiuser MIMO, where several users are treated by the BS as a single virtual user with multiple antennas [86]. Virtual MIMO is supported in the current LTE standard, but its implementation is transparent to the users involved [86]. When user cooperation is assumed, the topic of multiuser MIMO touches upon the subject of coalitional game theory.

14.3   COALITIONAL GAME THEORY

In coalitional game theory, users form coalitions and act as a virtual single user with respect to the BS. A user becomes a member of a coalition if its individual utility (e.g., rate) is greater within the coalition than when it acts individually [43, 46, 214]. This represents an important extension to the distributed scheduling schemes presented in this book. For example, users in a coalition can elect a representative user to communicate with the BS while communicating inside the coalition using the proportional fair distributed scheduling approach described in Chapter 9. The role of the elected user thus corresponds to the role of the CCD in Chapters 9 and 10. Resource allocation can in this case be subdivided into two phases. In the first phase, the resources are allocated by the BS to the coalitions, where each coalition can be seen as a single virtual user. A possible approach would be to follow the same principles used in the allocation of resources by the BS to the remote radio heads in a distributed base station scenario. In the second phase, the resources received from the BS are allocated to the users within each coalition based on a pre-defined utility function to maximize benefit. Possible allocation techniques can follow the methods of Chapters 9 and 10. However, applying proportional fair scheduling with user collaboration as in Chapter 9 gains a particular importance in this case since it corresponds to the Nash bargaining solution, which is the Pareto optimal solution in a bargaining problem involving the shared resources.

Thus, the two phases described above can easily be implemented using scheduling techniques presented in this book. The added novelty would be to discuss efficient protocols for coalition formation between users, and in selecting a representative of the coalition to communicate with the BS. If all users in a coalition are allowed to communicate with the BS while still appearing as a single entity, then other resource allocation algorithms need to be derived. In this case, virtual MIMO techniques involving several users might be needed. In all cases, the work in this book can be extended to a coalitional game theory framework by combining efficient resource allocation techniques with the coalitional game theory results obtained in Refs [43, 46, 214–219].

14.4   RESOURCE ALLOCATION WITH FEMTOCELLS

Resource allocation in the presence of femtocells was discussed in Sections 8.6 and 12.6.4. In these sections, an analysis of the applicability of the techniques presented in this book to resource allocation in the presence of femtocells was presented.

The use of femtocells is expected to improve the overall performance of the wireless users by providing dedicated BSs, the femtocell access points, at a closer distance to the users, thus ensuring a more efficient resource allocation both within the coverage of the femtocells and within the coverage of the external macrocell. In fact, femtocell access points lead to offloading traffic from the microcell/macrocell network. This leads to the availability of additional resources to the macrocell users, which allows these users to be allocated more subcarriers. Nonetheless, resource allocation in the presence of femtocells yields numerous technical challenges, notably due to the presence of interference between the femtocells and the main wireless network.

In fact, resource allocation in macrocell networks depends on the interference caused by the femtocell access points, and hence on the density of their deployment [141, 179, 180]. Power control techniques [179, 181], self-organizing frequency assignment schemes [182], and fractional frequency reuse [183–185], have been adopted to perform resource allocation with femtocells in order to provide interference avoidance for coexistence with macrocells.

Hence, the challenges of resource allocation in the presence of femtocells, the interplay of resource allocation between femtocells and traditional macrocells, and techniques for interference mitigation and/or coordination in the scenario of joint macrocell/femtocell deployments, are interesting topics for future investigation. Elaborate collaborative and noncollaborative techniques can be derived to reduce the mutual interference between neighboring femtocells, and between femtocells and the surrounding macrocell BSs.

14.5   GREEN NETWORKS AND SELF-ORGANIZING NETWORKS

Traditional efforts for saving energy in cellular networks focus on reducing the transmit powers of BSs. However, when a BS is in its working mode, studies show that more than 50% of the energy consumed is due to internal processing, air conditioning, and other factors [220]. In addition to the growing awareness of energy efficient wireless networks and their environmental benefits, electricity bills have become a significant cost factor for mobile operators [221]. This trend in reducing energy consumption and ensuring green wireless networks is supported by political and national initiatives that are beginning to put requirements on lowering the CO_2 emissions. For example, the European Commission research project EARTH concentrates on energy efficiency in radio access networks with the goal of finding solutions and concepts that can reduce energy consumption of mobile broadband systems by 50% [222].

An effective approach to perform resource allocation while saving energy in cellular networks is to completely turn off selected BSs when the traffic load becomes light, such that the BSs that are still on are sufficient to ensure the desired QoS for the existing users, for example, see Refs [223–227]. Switching off BSs leads to considerable energy savings, since traffic in peak hours can be as much as 10 times higher than the traffic in off-peak periods in the same area [221, 228]. To maintain a smooth network operation guaranteeing user satisfaction, enough resources should be available so that they can be allocated to the requesting users whenever the need arises. Thus, BSs that are turned off should know when to automatically switch on based on either local intelligent decisions, external trigger, or reconfiguration schedule. Hence, there is a need for intelligence among BSs in order to reconfigure the network automatically and dynamically based on current traffic situation, which leads to a form of self-organizing networks (SONs) [229].

Interesting research directions include evolved methods for actually reaching the scenario of green network deployments, beyond the current approach of switching BSs on and off intelligently. For example, in Ref. [226], cell zooming is proposed as an extension to the simple approach of turning on/off BS sites. Another interesting approach would be to integrate BS switching as part of the resource allocation process, where the energy consumed at the BS would be incorporated in the formulation of the resource allocation problem in addition to the transmit power of the BSs and of the mobile users. In this case, distributed pricing schemes or collaborative schemes can be developed in order to optimize the resource allocation performance with the additional constraint of achieving green operation.

14.6   JOINT UPLINK/DOWNLINK RESOURCE ALLOCATION

The increase in demand for delay-sensitive applications with symmetric high rate requirements, such as mobile gaming and video conferencing, has mandated the need for efficient resource allocation schemes in state-of-the-art OFDMA-based wireless communications systems. As resources for the uplink and downlink are usually separated, traditional approaches to resource allocation treated the uplink and downlink independently. In order to provide delay guarantees for real time services in wireless networks, queue aware resource allocation schemes have been recently proposed and studied for both the uplink and the downlink directions independently, for example, [230].

The problem objective and formulation change significantly when one considers the coupling between the uplink and downlink directions in determining the QoS of a given user who is running a symmetrical real time service, where good quality in both directions is required to achieve end user satisfaction. For instance, a user with a mobile gaming application having very good channel conditions in the uplink and bad channel conditions in the downlink will suffer from bad overall quality because it will be able to send with a high rate but receive with a very limited rate that notably affects service interactivity. Moreover, since the overall resources in the network are limited, it would be more beneficial to allocate that user's uplink resources to another user that might have worse uplink channel conditions but can achieve higher overall bidirectional QoS benefit. Hence, it is important to investigate joint uplink/downlink resource allocation schemes, especially for services with symmetrical rate and delay requirements.

A generalized framework for joint uplink/downlink resource allocation is laid down in Ref. [231] whereas some existing algorithms deal with asymmetrical services [232] or TDD–OFDMA based systems [233]. None of these references consider optimized resource allocation for symmetrical services with explicit modeling of queue states and delays as part of the problem formulation. As future work, one can formulate and solve the resource allocation problem in OFDMA networks by jointly considering both the uplink and the downlink directions. The problem formulation should take into account the uplink and downlink queue and channel states in order to perform optimized subcarrier and power allocation among users.

14.7 JOINT RESOURCE ALLOCATION IN HETEROGENEOUS NETWORKS

With the deployment of various types of wireless technologies, it is not uncommon to have different radio access technologies (RATs) covering the same areas, or having an intersection in their coverage of different areas. Furthermore, mobile devices capable of supporting multiple standards are becoming common in the market. Consequently, optimizing resource allocation in such heterogenous networks is an important topic for investigation. The heterogeneous networks can be based on OFDMA as a common access method, for example, the case of coexistence of LTE and WiMAX. However, they can be based on different access technologies, for example, WiMAX and WiFi. The distributed techniques presented in Chapters 9 and 10 could be used as novel concepts for OFDMA based LANs, thus allowing mobiles to use the same access method on the long-range macrocell network and on the short-range LAN. However, convergence over the air interface with OFDMA as the method of wireless access is not reached yet, and in practice, several technologies coexist. Hence, optimizing resource allocation over several RATs is a research direction of practical interest.

Inter-RAT handover is considered in the standards, where a user is handed over from a technology to the other based on its signal quality levels within the different RATs [234, 235]. However, the objective of this section goes beyond simple handover. The purpose is to perform joint resource allocation between the different RATs.

Inter-RAT resource allocation can be used to perform load balancing between the different RATs, thus reducing the load on an occasionally congested RAT. This allows to benefit from the presence of the different RATs in order to provide better QoS for the different users, and to provide these users with ubiquitous and seamless wireless access. Some research initiatives already started to investigate resource allocation in heterogenous networks, by proposing methods to select the best network to serve a user, among several available networks such as GSM/EDGE, UMTS/HSPA, WiMAX, and WLAN [50–52]. In Ref. [236], bandwidth sharing in a heterogeneous network consisting of WiMAX and WiFi is considered and a pricing model is derived for this purpose. In Ref. [237], a scenario where a WiMAX network is used as backup for a WiFi network failure is investigated.

However, optimized resource allocation in heterogeneous networks is not sufficiently studied. Designing frameworks for inter-RAT radio resource management in order to ensure user satisfaction in heterogeneous networks is a topic that should attract more research attention. In this case, operator revenue should be incorporated in the designed framework. In fact, the purpose of inter-RAT or cross-RAT resource allocation should not be at the expense of reducing operators' revenues. If the different RATs are managed by the same operator, then resource allocation in heterogeneous networks will increase the operator's revenue since it will lead to a reduction in user drop rate, an increase in user satisfaction, and an optimized usage of the wireless resources of the various RATs.

In the case where different operators run the different RATs, an inter-operator pricing/billing model should be incorporated in the optimization framework as an additional parameter affecting the inter-RAT handover decisions. In this case, operator revenues should be taken into account before taking a decision of moving a user from an operator's network to another operator's network. For example, optimizing load balancing over the networks of the different operators would not be an absolute necessity; however, avoiding to drop a user by allowing an interoperator handover would increase user satisfaction without decreasing the operator's revenue. These concepts also apply to the case of interoperator handover/resource allocation over the same RAT (not necessarily different RATs), for example, in the scenario of two operators each running its own LTE network with overlapping coverage.

14.8 RESOURCE ALLOCATION IN COGNITIVE RADIO NETWORKS

Cognitive radio (CR) networks have gained increasing research importance, and the problem of resource allocation in CR networks is being widely investigated in the recent literature [61, 238–240]. In this book, topics related to CR were investigated in different chapters, especially in Sections 10.7.6 and 12.6.3, where indications on research directions relevant to CR were given. In this section, more insights on resource allocation in CR networks are presented.

In CR networks, secondary or unlicensed users benefit opportunistically from the spectrum while protecting primary or licensed users from interference [238].

Three techniques are used to refer to the solution adopted in a CR scenario: underlay, overlay, and interweave [241]. In the "underlay" technique, simultaneous primary and secondary transmissions are allowed. Secondary users spread their signal over a bandwidth large enough to ensure that the amount of interference caused to the primary users is within tolerable limits. In the "overlay" technique, simultaneous primary and secondary transmissions are also allowed, but secondary users are assumed aware of the primary message. They use this side information to mitigate the interference caused by primary users, in addition to relaying the primary signal, thus compensating the interference caused to primary users by their simultaneous transmission. The "interweave" technique is based on opportunistic transmission, where secondary users monitor the radio spectrum, detect the presence/absence of primary users in the different frequency bands and then transmit their secondary signals in the vacant frequency slots. This technique requires accurate detection of the presence of primary users, and a tradeoff between sensing time and transmission time is needed, for example, [61, 239, 240]. In some references, the CR classification is performed into only overlay and underlay, with the term overlay in this case given the same definition as the interweave scenario [238, 242].

In general, underlay CR networks are investigated in the context of ultra-wide band (UWB) transmission in order to ensure low secondary transmit power and hence low interference to primary users, for example, [242]. In Ref. [238], resource allocation in CDMA underlay CR networks is considered, with interference constraints for primary users and quality of service constraints for secondary users. In Ref. [243], a cognitive radio network pricing model is used in order to admit new users to an access network by service providers while minimizing the effect on QoS of existing users.

With OFDMA adopted for state-of-the-art wireless communication systems, research on OFDMA CR networks received increasing attention. In fact, OFDMA is used as the accessing scheme of CR networks in IEEE 802.22 [244]. In Ref. [33], ergodic resource allocation in CR OFDMA networks is considered. The problem is formulated as a joint utility maximization problem for both primary and secondary users, with minimum rate constraints imposed to ensure QoS for primary users. All users in the network (primary and secondary) are assumed to exchange information in order to perform this maximization in a distributed way.

However, most of the existing CR literature neglect the interference from the primary to the secondary network, or consider simplified models in their derivations. In addition, mobile devices are assumed to be equipped with two antennas in most state-of-the-art OFDMA-based wireless communication systems [245]. This property has been seldom exploited in resource allocation in CR networks. In Ref. [246], the availability of two transmit antennas at the primary and secondary mobile users is exploited by treating these antennas as a two-element antenna array. Secondary users use this antenna array to place a null in the array pattern in the direction of the primary BS. This leads to a protection of primary users from secondary interference, since in the uplink the interference is measured at the BS and is not dependent on the user location as in the downlink. The primary users benefit from their two-element antenna array in order to perform beam steering in the direction of the primary BS. This will lead to enhanced SINR and hence a higher rate for primary users since the radiation is concentrated in the direction of the BS. This approach has the unintentional side effect of reducing the transmitted radiation in other directions, and hence leads to an interference reduction in the secondary network, although primary users are not concerned with the presence of secondary users and hence there is no exchange of information between the primary and secondary networks.

However, more research is needed in the area of OFDMA CR networks, with general formulations taking into account interference, optimized resource allocation, and QoS constraints for both primary and secondary users. Therefore, resource allocation in OFDMA CR networks still constitutes an interesting topic for future research.

BIBLIOGRAPHY

1. T. S. Rappaport, *Wireless Communications: Principles and Practice*, 2nd edition, Prentice Hall, 2002.

2. A. Goldsmith, *Wireless Communications*, Cambridge University Press, 2005.

3. GSM World, Market Data Summary, Online—last viewed: December 07, 2009, http://www.gsmworld.com/newsroom/market-data/.

4. UMTS Forum, Fast Facts, Online—last viewed: December 07, 2009, http://www.umts-forum.org/.

5. H. Holma and A. Toskala, *WCDMA for UMTS: Radio Access for Third Generation Mobile Communications*, 3rd edition, Wiley, 2004.

6. J. G. Andrews, A. Ghosh, and R. Muhamed, *Fundamentals of WiMAX*, Pearson Education, 2007.

7. R. W. Chang, Synthesis of Band-Limited Orthogonal Signals for Multichannel Data Transmission, *Bell Systems Technical Journal*, 45, 1775–1796, 1966.

8. B. R. Saltzberg, Performance of an Efficient Parallel Data Transmission System, *IEEE Transactions on Communications Technology*, 15(6), 805–811, 1967.

9. S. Pietrzyk, *OFDMA for Broadband Wireless Access*, Artech House, 2006.

10. S. B. Weinstein and P. M. Ebert, Data Transmission by Frequency-Division Multiplexing Using the Discrete Fourier Transform, *IEEE Transactions on Communications Technology*, 19(5), 628–634, 1971.

11. P. S. Chow, J. C. Tu, and J. M. Cioffi, A Discrete Multitone Transceiver System for HDSL Applications, *IEEE Journal on Selected Areas in Communications*, 9(6), 895–908, 2001.

12. B. R. Saltzberg, Comparison of Single-Carrier and Multitone Digital Modulations for ADSL Applications, *IEEE Communications Magazine*, 36(11), 114–121, 1998.

13. T. Starr, J. M. Cioffi, and P. Silverman, *Understanding Digital Subscriber Line Technology*, Prentice Hall, 1998.

14. IEEE 802.11, Wireless LAN Medium Access Control (MAC) and Physical (PHY) Layer Specifications, 1999.

15. European Telecommunications Standards Institute (ETSI), Digital Video Broadcasting: Framing Structure, Channel Coding and Modulation for Digital Terrestrial Television, ETSI EN 300-744, 1997.

16. T. Keller and L. Hanzo, Adaptive Multicarrier Modulation: A Convenient Framework for Time-Frequency Processing in Wireless Communications, *Proceedings of the IEEE*, 88(5), 611–640, 2000.

Resource Allocation in Uplink OFDMA Wireless Systems: Optimal Solutions and Practical Implementations, Elias E. Yaacoub and Zaher Dawy.
© 2012 by the Institute of Electrical and Electronics Engineers, Inc. Published 2012 by John Wiley & Sons, Inc.

17. I. Kalet, The Multitone Channel, *IEEE Transactions on Communications*, 37(2), 119–124, 1989.

18. C. Y. Wong, R. S. Cheng, K. B. Letaief, and R. D. Murch, Multiuser OFDM with Adaptive Subcarrier, Bit and Power Allocation, *IEEE Journal on Selected Areas in Communications*, 17(10), 1747–1757, 1999.

19. G. Song and Y. Li, Cross-Layer Optimization for OFDM Wireless Networks—Part I: Theoretical Framework, *IEEE Transactions on Wireless Communications*, 4(2), 614–624, 2005.

20. G. Song and Y. Li, Cross-Layer Optimization for OFDM Wireless Networks—Part II: Algorithm Development, *IEEE Transactions on Wireless Communications*, 4(2), 625–634, 2005.

21. C.Y. Wong, C.Y. Tsui, R.S. Cheng, and K.B. Letaief, A Real-time Sub-carrier Allocation Scheme for Multiple Access Downlink OFDM Transmission, *IEEE VTC-Fall 1999*, September 1999.

22. L.-C. Wang and W.-J. Lin, Throughput and Fairness Enhancement for OFDMA Broadband Wireless Access Systems Using the Maximum *C/I* Scheduling, *IEEE VTC-Fall 2004*, September 2004.

23. G. Song, Y. Li, L. J. Cimini, Jr., and H. Zheng, Joint Channel-Aware and Queue-Aware Data Scheduling in Multiple Shared Wireless Channels, *IEEE WCNC 2004*, March 2004.

24. G. Song and Y. Li, Adaptive Subcarrier and Power Allocation in OFDM Based on Maximizing Utility, *IEEE VTC-Spring 2003*, vol. 2, April 2003, pp. 905–909.

25. S. Pietrzyk and G.J.M. Janssen, Multiuser Subcarrier Allocation for QoS Provision in the OFDMA Systems, *IEEE VTC-Fall 2002*, September 2002.

26. I. C. Wong and B. L. Evans, Optimal Downlink OFDMA Resource Allocation with Linear Complexity to Maximize Ergodic Rates, *IEEE Transactions on Wireless Communications*, 7(3), 962–971, 2008.

27. X. Wang and G. B. Giannakis, Ergodic Capacity and Average Rate-Guaranteed Scheduling for Wireless Multiuser OFDM Systems, *IEEE ISIT 2008*, July 2008.

28. R. Knopp, P.A. Humblet, Information Capacity and Power Control in Single-Cell Multiuser Communications, *IEEE ICC 1995*, June 1995.

29. P. Viswanath, D.N.C. Tse, R. Laroia, Opportunistic Beamforming Using Dumb Antennas, *IEEE Transactions on Information Theory*, 48(6), 1277–1294, 2002.

30. S. Pfletschinger, G. Münz, and J. Speidel, Efficient Subcarrier Allocation for Multiple Access in OFDM Systems, *International OFDM-Workshop 2002*, September 2002.

31. I. C. Wong and B. L. Evans, Optimal OFDMA Resource Allocation with Linear Complexity to Maximize Ergodic Weighted Sum Capacity, *IEEE Internation Conference on Acoustics, Speech, and Signal Proceedings.*, 2, April 2007, pp. 601–604.

32. A. M. El-Hajj, E. Yaacoub, and Z. Dawy, Uplink OFDMA Resource Allocation with Ergodic Rate Maximization, *IEEE PIMRC 2009*, September 2009.

33. J.-A. Bazerque and G. B. Giannakis, Distributed Scheduling and Resource Allocation for Cognitive OFDMA Radios, *Mobile Networks and Applications*, 13(5), 452–462, 2008.

34. S. Boyd and L. Vandenberghe, *Convex Optimization*, Cambridge University Press, 2004.

35. D. P. Bertsekas, *Nonlinear Programming*, 2nd edition, Athena Scientific, 1999.

36. J.-A. Bazerque and G. B. Giannakis, Distributed Scheduling and Resource Allocation for Cognitive OFDMA Radios, *International Conference on Cognitive Radio Oriented Wireless Networks and Communications*, August 2007.

37. D. N. C. Tse and S. V. Hanly, Multiaccess Fading Channels—Part I: Polymatroid Structure, Optimal Resource Allocation and Throughput Capacities, *IEEE Transactions on Information Theory*, 44(7), 2796–2815, 1998.

38. T.M. Cover and J.A. Thomas, *Elements of Information Theory*, 2nd edition, John Wiley and Sons, 2006.

39. D. Kivanc, G. li, and H. Liu, Computationally Efficient Bandwidth Allocation and Power Control for OFDMA, *IEEE Transactions on Wireless Communications*, 2(6), 1150–1158, 2003.

40. H. Yin and H. Liu, An Efficient Multiuser Loading Algorithm for OFDM-Based Broadband Wireless Systems, *IEEE GlobeCom 2000*, December 2000.

41. T. Lunttila, J. Lindholm, K. Pajukoski, E. Tiirola, and A. Toskala, EUTRAN Uplink Performance, *International Symposium on Wireless Pervasive Computing (ISWPC) 2007*, February 2007.

42. IEEE 802.16-2004, IEEE Standard for Local and Metropolitan Area Networks, Air Interface for Fixed Broadband Wireless Access Systems, 2004.

43. H. Boche and M. Schubert, Nash Bargaining and Proportional Fairness for Wireless Systems, *IEEE/ACM Transactions on Networking*, 17(5), 1453–1466, 2009.

44. J. F. Nash, Jr., The Bargaining Problem, *Econometrica*, 18, 155–162, 1950.

45. F. Carmichael, *A Guide to Game Theory*, Prentice Hall, 2005.

46. S. Mathur, L. Sankaranarayanan, and N. B. Mandayam, Coalitional Games in Gaussian Interference Channels, *IEEE ISIT 2006*, July 2006.

47. C. Y. Ng and C. W. Sung, Low Complexity Subcarrier and Power Allocation for Utility Maximization in Uplink OFDMA Systems, *IEEE Transactions on Wireless Communications*, 7(5), 1667–1675, 2008.

48. K. C. Beh, S. Armour, and A. Doufexi, Joint Time-Frequency Domain Proportional Fair Scheduler with HARQ for 3GPP LTE Systems, *IEEE VTC 2008-Fall*, September 2008.

49. J. Choi, H. Lee, H. Chung, and J. H. Lee, Sounding Method for Proportional Fair Scheduling in OFDMA/FDD Uplink, *IEEE VTC 2007-Spring*, April 2007.

50. J. Bühler and G. Wunder, An Optimization Framework for Heterogeneous Access Management, *IEEE WCNC 2009*, April 2009.

51. Y. Wang, L. Zheng, J. Yuan, and W. Sun, Median Based Network Selection in Heterogeneous Wireless Networks, *IEEE WCNC 2009*, April 2009.

52. W. Luo and E. Bodanese, Radio Access Network Selection in a Heterogeneous Communication Environment, *IEEE WCNC 2009*, April 2009.

53. J. Huang, R. A. Berry, and M. L. Honig, Distributed Interference Compensation for Wireless Networks, *IEEE Journal on Selected Areas in Communications*, 25(5), 1074–1084, 2006.

54. Q. Zhang, J. Jia, and J. Zhang, Cooperative Relay to Improve Diversity in Cognitive Radio Networks, *IEEE Communications Magazine*, 47(2), 111–117, 2009.

55. H. Shan, W. Zhuang, and Z. Wang, Distributed Cooperative MAC for Multihop Wireless Networks, *IEEE Communications Magazine*, 47(2), 126–133, 2009.

56. H. Jiang, P. Wang, and W. Zhuang, A Distributed Channel Access Scheme with Guaranteed Priority and Enhanced Fairness, *IEEE Transactions on Wireless Communications*, 6(6), 2114–2125, 2007.

57. P. Wang and W. Zhuang, A Token-Based Scheduling Scheme for WLANs Supporting Voice/Data Traffic and Its Performance Analysis, *IEEE Transactions on Wireless Communications*, 7(5), 1708–1718, 2008.

58. P. Zhu, J. Li, and X. Wang, Scheduling Model for Cognitive Radio, *International Conference on Cognitive Radio Oriented Wireless Networks and Communications*, May 2008.

59. M. Thoppian, S. Venkatesan, R. Prakash, and R. Chandrasekaran, MAC-layer Scheduling in Cognitive Radio based Multi-hop Wireless Networks, *International Symposium on a World of Wireless, Mobile and Multimedia Networks 2006 (WoWMoM'06)*, June 2006.

60. C. Peng, H. Zheng, and B. Y. Zhao, Utilization and Fairness in Spectrum Assignment for Opportunistic Spectrum Access, *Mobile Networks and Applications*, 11(4), 555–576, 2006.

61. A. T. Hoang and Y.-C. Liang, Adaptive Scheduling of Spectrum Sensing Periods in Cognitive Radio Networks, *IEEE GlobeCom 2007*, November 2007.

62. L. Cao and H. Zheng, Distributed Rule-Regulated Spectrum Sharing, *IEEE Journal on Selected Areas in Communications*, 26(1), 130–145, 2008.

63. S. Gandhi, C. Buragohain, L. Cao, H. Zheng, S. Suri, Towards Real-Time Dynamic Spectrum Auctions, *The International Journal of Computer and Telecommunications Networking*, 52(4), 879–897, 2008.

64. QualComm, 1xEV: 1x evolution, IS-856 TIA/EIA Standard, technical report, QualComm, 2004.

65. 3rd Generation Partnership Project 2 (3GPP2), cdma2000 High Rate Packet Data Air Interface Specification Revision A, Technical report c.s20024-a, March 2004.

66. CPRI Specification V2.0, Common Public Radio Interface (CPRI) Interface Specification, technical report, October 2004.

67. A. A. M. Saleh, A. J. Rustako, and R. S. Roman, Distributed Antennas for Indoor Radio Communications, *IEEE Transactions on Communications*, 35(12), 1245–1251, 1987.

68. R. E. Schuh and M. Sommer, WCDMA Coverage and Capacity Analysis for Active and Passive Distributed Antenna Systems, *IEEE VTC-Spring 2002*, May 2002.

69. L. Dai, S. Zhou, and Y. Yao, Capacity with MRC-based Macrodiversity in CDMA Distributed Antenna Systems, *IEEE GlobeCom 2002*, November 2002.

70. L. Dai, S. Zhou, and Y. Yao, Capacity Analysis in CDMA Distributed Antenna Systems, *IEEE Transactions on Wireless Communications*, 4(6), 2613–2620, 2005.

71. R. Hasegawa, M. Shirakabe, R. Esmailzadeh, and M. Nakagawa, Downlink Performance of a CDMA System with Distributed Base Station, *IEEE VTC-Fall 2003*, October 2003.

72. W. Choi, J. G. Andrews, and C. Yi, The Capacity of Multicellular Distributed Antenna Networks, *International Conference on Wireless Networks, Communications and Mobile Computing*, June 2005.

73. W. Choi and J. G. Andrews, Downlink Performance and Capacity of Distributed Antenna Systems in a Multicell Environment, *IEEE Transactions on Wireless Communications*, 6(1), 69–73, 2007.

74. J. Zhang and J. G. Andrews, Cellular Communication with Randomly Placed Distributed Antennas, *IEEE GlobeCom 2007*, November 2007.

75. J. Zhang and J. G. Andrews, Distributed Antenna Systems with Randomness, *IEEE Transactions on Wireless Communications*, 7(9), 3636–3646, 2008.

76. H. Fujii and H. Yoshino, Theoretical Capacity and Outage Rate of OFDMA Cellular System with Fractional Frequency Reuse, *IEEE VTC-Spring 2008*, May 2008.

77. R. Kwan and C. Leung, A Survey of Scheduling and Interference Mitigation in LTE, *Journal of Electrical and Computer Engineering*, Article ID 273486, doi:10.1155/2010/273486, vol. 2010, 2010.

78. Z. Xie and B. Walke, Performance Analysis of Reuse Partitioning Techniques in OFDMA based Cellular Radio Networks, *International Conference on Telecommunications (ICT 2010)*, April 2010.

79. H. Zhang, L. Venturino, N. Prasad, and S. Rangarajan, Distributed Inter-Cell Interference Mitigation in OFDMA Wireless Data Networks, *IEEE GlobeCom Workshops 2008*, December 2008.

80. A. Gjendemsjø, D. Gesbert, G. E. Øien, and S. G. Kiani, Binary Power Control for Sum Rate Maximization over Multiple Interfering Links, *IEEE Transactions on Wireless Communications*, 7(8), 3164–3173, 2008.

81. S. G. Kiani, G. E. Øien, and D. Gesbert, Maximizing Multicell Capacity Using Distributed Power Allocation and Scheduling, *IEEE WCNC 2007*, March 2007.

82. Z. Liang, Y. H. Chew, and C. C. Ko, Decentralized Bit, Subcarrier and Power Allocation with Interference Avoidance in Multicell OFDMA Systems using Game Theoretic Approach, *IEEE MILCOM 2008*, November 2008.

83. 3rd Generation Partnership Project (3GPP), 3GPP TS 36.213 3GPP TSG RAN Evolved Universal Terrestrial Radio Access (E-UTRA) Physical layer procedures, version 8.3.0, Release 8, 2008.

84. 3rd Generation Partnership Project (3GPP), 3GPP TR 36.942 3GPP TSG RAN Evolved Universal Terrestrial Radio Access (E-UTRA) Radio Frequency (RF) system scenarios, version 8.1.0, Release 8, 2008.

85. 3rd Generation Partnership Project (3GPP), 3GPP TS 36.423 3GPP TSG RAN Evolved Universal Terrestrial Radio Access (E-UTRA) X2 application protocol (X2AP), version 9.3.0, Release 9, 2010.

86. G. Boudreau, J. Panicker, N. Guo, R. Chang, N. Wang, and S. Vrzic, Interference Coordination and Cancellation for 4G Networks, *IEEE Communications Magazine*, 47(4), 74–81, 2009.

87. G. Fodor, C. Koutsimanis, A. Rácz, N. Reider, A. Simonsson, and W. Müller, Intercell Interference Coordination in OFDMA Networks and in the 3GPP Long Term Evolution System, *Journal of Communications*, 4(7), 445–453, 2009.

88. J. Ellenbeck, M. Hammoud, B. Lazarov, and C. Hartmann, Autonomous Beam Coordination for the Downlink of an IMT-Advanced Cellular System, *European Wireless (EW 2010)*, April 2010.

89. R. Irmer et al., Multisite Field Trial for LTE and Advanced Concepts, *IEEE Communications Magazine*, 47(2), 92–98, 2009.

90. I. Akyildiz, D. Gutierrez-Estevez, and E. Reyes, The Evolution to 4G Cellular Systems: LTE-Advanced, *Physical Communication*, 3(4), 217–244, 2010.

91. M. Boldi, E. Hardouin, M. Olsson, H. Pennanen, and A. Tölli, Coordinated Beamforming for IMT-Advanced in the Framework of WINNER+ Project, *Future Network and Mobile Summit*, June 2010.

92. L. Thiele, F. Boccardi, C. Botella, T. Svensson, and M. Boldi, Scheduling-Assisted Joint Processing for CoMP in the Framework of the WINNER+ Project, *Future Network and Mobile Summit*, June 2010.

93. WINNER+, Celtic Project CP5-026 Intermediate Report on CoMP (Coordinated Multi-Point) and Relaying in the Framework of CoMP (S. Mayrargue, ed.), Public Deliverable D1.8, Wireless World Initiative New Radio - WINNER+, http://projects.celtic-initiative.org/winner+/index.html, November 2009.

94. A. Osseiran, J. Monserrat, and W. Mohr, *Mobile and Wireless Communications for IMT—Advanced and Beyond*, John Wiley and Sons, July 2011.

95. K. Kim, Y. Han, and S.-L. Kim, Joint Subcarrier and Power Allocation in Uplink OFDMA Systems, *IEEE Communications Letters*, 9(6), 526–528, 2005.

96. L. Gao and S. Cui, Efficient Subcarrier, Power, and Rate Allocation with Fairness Consideration for OFDMA Uplink, *IEEE Transactions on Wireless Communications*, 7(5), 1507–1511, 2008.

97. J. Huang, V. G. Subramanian, R. Agrawal, and R. Berry, Joint Scheduling and Resource Allocation in Uplink OFDM Systems for Broadband Wireless Access Networks, *IEEE Journal on Selected Areas in Communications*, 27(2), 226–234, 2009.

98. 3rd Generation Partnership Project (3GPP), 3GPP TR 25.814 3GPP TSG RAN Physical Layer Aspects For Evolved UTRA, v7.1.0, 2006.

99. A. Alsawah and I. Fijalkow, Weighted Sum-Rate Maximization in Multiuser-OFDM Systems under Differentiated Quality-of-Service Constraints, *IEEE 8th Workshop on Signal Processing Advances in Wireless Communications (SPAWC) 2007*, June 2007.

100. S. Pfletschinger, Achievable Rate Regions for OFDMA with Link Adaptation, *IEEE PIMRC 2007*, September 2007.

101. L. Li and A. J. Goldsmith, Optimal Resource Allocation for Fading Broadcast Channels—Part I: Ergodic Capacity, *IEEE Transactions on Information Theory*, 47(3), 1083–1102, 2001.

102. A. J. Goldsmith and M. Effros, The Capacity Region of Broadcast Channels with Intersymbol Interference and Colored Gaussian Noise, *IEEE Transactions on Information Theory*, 47(1), 219–240, 2001.

103. W. Yu and J. M. Cioffi, FDMA Capacity of Gaussian Multiple-Access Channels with ISI, *IEEE Transactions on Communications*, 50(1), 102–111, 2002.

104. L. M. Hoo, B. Halder, J. Tellado, and J. M. Cioffi, Multiuser Transmit Optimization for Multicarrier Broadcast Channels: Asymptotic FDMA Capacity Region and Algorithms, *IEEE Transactions on Communications*, 52(6), 922–930, 2004.

105. B. Ghimire, G. Auer, and H. Haas, OFDMA-TDD Networks with Busy Burst Enabled Grid-of-Beam Selection, *IEEE ICC 2009*, June 2009.

106. A. Tölli, H. Pennanen, and P. Komulainen, Distributed Coordinated Multi-Cell Transmission Based on Dual Decomposition, *IEEE GlobeCom 2009*, December 2009.

107. A. Tölli, H. Pennanen, and P. Komulainen, Decentralized Minimum Power Multi-Cell Beamforming with Limited Backhaul Signaling, *IEEE Transactions on Wireless Communications*, 10(2), 570–580, 2011.

108. A. Prékopa, K. Yoda, and M. M. Subasi, Uniform Quasi-Concavity in Probabilistic Constrained Stochastic Programming, Rutcor Research Report RRR 6-2010, RUTCOR, April 2010.

109. S. T. Chung and A. Goldsmith, Degrees of Freedom in Adaptive Modulation: A Unified View, *IEEE Transactions on Communications*, 49(9), 1561–1571, 2001.

110. W. Yu and J. M. Cioffi, Constant-Power Waterfilling: Performance Bound and Low-Complexity Implementation, *IEEE Transactions on Communications*, 54(1), 23–28, 2006.

111. J. Lim, H.G. Myung, K. Oh, and D.J. Goodman, Channel-Dependent Scheduling of Uplink Single Carrier FDMA Systems, *IEEE VTC-Fall 2006*, September 2006.

112. X. Qiu and K. Chawla, On the Performance of Adaptive Modulation in Cellular Systems, *IEEE Transactions on Communications*, 47(6), 884–895, 1999.

113. V. Tarokh, A. Naguib, N. Seshadri, and A. R. Calderbank, Space-Time Codes for High Data Rate Wireless Communication: Performance Criteria in the Presence of Channel Estimation Errors, Mobility, and Multiple Paths, *IEEE Transactions on Communications*, 47(2), 199–207, 1999.

114. Y. Chen and C. Tellambura, Performance Analysis of Maximum Ratio Transmission with Imperfect Channel Estimation, *IEEE Communications Letters*, 9(4), 322–324, 2005.

115. Y. Chen and N. C. Beaulieu, Maximum Likelihood Receivers for Space-Time Coded MIMO Systems with Gaussian Estimation Errors, *IEEE Transactions on Communications*, 57(6), 1712–1720, 2009.

116. 3rd Generation Partnership Project (3GPP), 3GPP TS 36.211 3GPP TSG RAN Evolved Universal Terrestrial Radio Access (E-UTRA) Physical Channels and Modulation, version 8.3.0, Release 8, 2008.

117. J. Lim, H.G. Myung, K. Oh, and D.J. Goodman, Proportional Fair Scheduling of Uplink Single-Carrier FDMA Systems, *IEEE PIMRC 2006*, September 2006.

118. F. P. Kelly, A. K. Maulloo, and D. K. H. Tan, Rate Control for Communication Networks: Shadow Prices, Proportional Fairness and Stability, *Journal of Operational Research Society*, 49, 237–252, 1998.

119. J. M. Holtzman, Asymptotic Analysis of Proportional Fair Algorithm, *IEEE PIMRC 2001*, October 2001.

120. G. Fodor, A. Furuskär, P. Skillermark, and J. Yang, On the Impact of Uplink Scheduling on Intercell Interference Variation in MIMO OFDM Systems, *IEEE WCNC 2009*, April 2009.

121. K. Kim, H. Kim, and Y. Han, Subcarrier and Power Allocation in OFDMA Systems, *IEEE VTC-Fall 2004*, September 2004.

122. M. V. Clark, T. M. Willis, L. J. Greenstein, A. J. Rustako, V. Ercegt, and R. S. Roman, Distributed versus Centralized Antenna Arrays in Broadband Wireless Networks, *IEEE VTC-Spring 2001*, May 2001.

123. W. Roh and A. Paulraj, Outage Performance of the Distributed Antenna Systems in a Composite Fading Channel, *IEEE VTC-Fall 2002*, September 2002.

124. Y. Hadisusanto, L. Thiele, and V. Jungnickel, Distributed Base Station Cooperation via Block-Diagonalization and Dual-Decomposition, *IEEE GlobeCom 2008*, December 2008.

125. P. Leroux, S. Roy, and J.-Y. Chouinard, A Multi-Agent Protocol to Manage Interference in a Distributed Base Station System, *International Conference on Advanced Technologies for Communications (ATC 2008)*, October 2008.

126. W. R. Highsmith, An Investigation into Distributed Base Station Design for LMDS Systems, *Proceedings IEEE Southeastcon 2002*, 2002.

127. Alcatel-Lucent, Alcatel-Lucent expands 3G W-CDMA/HSPA portfolio with new distributed base station that offers increased deployment flexibility and lowers power require-

ments, http://www.alcatel-lucent.com/wps/portal/NewsReleases/, Alcatel-Lucent, March 2008.

128. Alcatel-Lucent, Alcatel-Lucent unveils 3G CDMA/EV-DO distributed base station that offers greater deployment flexibility while lowering power requirements, http://www.alcatel-lucent.com/wps/portal/NewsReleases/, Alcatel-Lucent, April 2008.

129. H. Yanikomeroglu, Fixed and Mobile Relaying Technologies for Cellular Networks, *Second Workshop on Applications and Services in Wireless Networks (ASWN'02)*, July 2002, pp. 75–81.

130. R. Pabst, B. Walke, D. Schultz, P. Herhold, H. Yanikomeroglu, S. Mukherjee, H. Viswanathan, M. Lott, W. Zirwas, M. Dohler, H. Aghvami, D. Falconer, and G. Fettweis, Relay-Based Deployment Concepts for Wireless and Mobile Broadband Cellular Radio, *IEEE Communications Magazine*, 42(9), 80–89, 2004.

131. V. Sreng, H. Yanikomeroglu, and D. Falconer, Capacity Enhancement Through Two-Hop Relaying in Cellular Radio Systems, *IEEE WCNC 2002*, March 2002, pp. 881–885.

132. M. Salem, A. Adinoyi, M. Rahman, H. Yanikomeroglu, D. Falconer, Y.-D. Kim, E. Kim, and Y.-C. Cheong, An Overview of Radio Resource Management in Relay-Enhanced OFDMA-Based Networks, *IEEE Communications Surveys and Tutorials*, 12(3), 422–438, 2010.

133. M. Kaneko and P. Popovski, Radio Resource Allocation Algorithm for Relay-Aided Cellular OFDMA System, *IEEE ICC 2007*, June 2007.

134. W. Nam, W. Chang, S.-Y Chung, and Y. Lee, Transmit Optimization for Relay-Based Cellular OFDMA Systems, *IEEE ICC 2007*, June 2007, pp. 75–81.

135. G. Li and H. Liu, Resource Allocation for OFDMA Relay Networks with Fairness Constraints, *IEEE Journal on Selected Areas in Communications*, 24(11), 2061–2069, 2006.

136. C. Bae and D.-H. Cho, Fairness-Aware Adaptive Resource Allocation Scheme in Multi-Hop OFDMA Systems, *IEEE Communications Letters*, 11(2), 134–136, 2007.

137. C. Bae and D.-H. Cho, Adaptive Resource Allocation Based on Channel Information in Multi-Hop OFDM Systems, *IEEE VTC-Fall 2006*, September 2006.

138. J. Lee, S. Park, H. Wang, and D. Hong, QoS-Guarantee Transmission Scheme Selection for OFDMA Multi-Hop Cellular Networks, *IEEE ICC 2007*, June 2007.

139. O. Oyman, Opportunistic Scheduling and Spectrum Reuse in Relay-Based Cellular OFDMA Networks, *IEEE GlobeCom 2007*, November 2007.

140. M. Kim and H. Lee, Radio Resource Management for a Two-Hop OFDMA Relay System in Downlink, *IEEE ISCC 2007*, July 2007.

141. V. Chandrasekhar and J. G. Andrews, Uplink Capacity and Interference Avoidance for Two-Tier Femtocell Networks, *IEEE Transactions on Wireless Communications*, 8(7), 3498–3509, 2009.

142. J. Hoydis and M. Debbah, Green, Cost-effective, Flexible, Small Cell Networks, *IEEE ComSoc MMTC E-Letter Special Issue on Multimedia Over Femto Cells*, 5(5), 2010.

143. V. Chandrasekhar, J. G. Andrews, and A. Gatherer, Femtocell Networks: A Survey, *IEEE Communications Magazine*, 46, 59–67, 2008.

144. J. Zhang and G. de la Roche, *Femtocells: Technologies and Deployment*, John Wiley & Sons, 2010.

145. S.-P. Yeh, S. Talwar, S.-C. Lee, and H. Kim, WiMAX Femtocells: A Perspective on Network Architecture, Capacity, and Coverage, *IEEE Communications Magazine*, 46, 58–65, 2008.

146. Y. Xi and E. M. Yeh, Equilibria and Price of Anarchy in Parallel Relay Networks with Node Pricing, *42nd Annual Conference on Information Sciences and Systems (CISS 2008)*, March 2008.

147. J. Musacchio and S. Wu, The Price of Anarchy in Competing Differentiated Services Networks, *46th Annual Allerton Conference*, September 2008.

148. 3rd Generation Partnership Project (3GPP), 3GPP TS 25.101 UMTS User Equipment (UE) radio transmission and reception (FDD), version 8.1.0, Release 8, 2008.

149. A. Sridharan, R. Subarraman, and R. Guérin, Distributed Uplink Scheduling in CDMA Networks, *Proceedings of IFIP-TC6 Networking Conference*, May 2007.

150. A. Sridharan, R. Subarraman, and R. Guérin, Distributed Uplink Scheduling in CDMA Systems, Research Report RR06-ATL12070139, Sprint ATL, 2006.

151. 3rd Generation Partnership Project 2 (3GPP2), cdma2000 High Rate Packet Data Air Interface Specification, Technical report c.s20024 v2.0, October 2000.

152. C. Lott, N. Bhushan, D. Ghosh, R. Attar, J. Au, and M. Fan, Reverse Traffic Channel MAC Design of cdma2000 1xEV-DO Revision A System, *IEEE VTC-Spring 2005*, May–June 2005.

153. S. Chakravarty, R. Pankaj, E. Esteves, An Algorithm for Reverse Traffic Channel Rate Control for cdma2000 High Rate Packet Data Systems, *IEEE GlobeCom 2001*, December 2001.

154. W. Zhongwei, Research on Uplink Rate Control Algorithm in UMTS, Technical report, Beijing University of Posts and Telecommunications, 2006.

155. E. Yaacoub, Z. Dawy, K. Y. Kabalan, and A. El-Hajj, Reverse Link Rate Control in 1xEV-DO with Adaptive Antenna Arrays, *IEEE International Wireless Communications and Mobile Computing Conference (IWCMC 2008)*, August 2008.

156. E. Yaacoub, Z. Dawy, A. El-Hajj, and K. Y. Kabalan, Distributed On-Off Uplink Scheduling in CDMA Systems with Adaptive Antenna Arrays, *IFIP Wireless Days Conference*, November 2008.

157. E. Yaacoub, Z. Dawy, A. El-Hajj, and K. Y. Kabalan, Distributed Uplink Scheduling and Rate Control in cdma2000 using Adaptive Antenna Arrays, *International Journal of Electronics and Communication (AEU)*, 63(10), 841–852, 2009.

158. IEEE 802.15.4, Wireless Medium Access Control (MAC) and Physical Layer (PHY) Specifications for Low-Rate Wireless Personal Area Networks (WPANs), 2006.

159. European Telecommunications Standards Institute (ETSI), Radio Equipment and Systems (RES), High Performance Radio Local Area Network (HIPERLAN) Type 1, Functional Specification, ETSI ETS 300 652, 1996.

160. J. H. Schiller, *Mobile Communications*, 2nd edition, Addison-Wesley, 2003.

161. F. Halsall, *Data Communications, Computer Networks, and Open Systems*, Addison-Wesley, 1996.

162. N. Abramson, The Throughput of Packet Broadcasting Channels, *IEEE Transactions on Communications*, COM-25(1), 117–128, 1977.

163. A. Salkintzis, Packet Data over Cellular Networks: The CDPD Approach, *IEEE Communications Magazine*, 37(6), 152–159, 1999.

164. European Telecommunications Standards Institute (ETSI), Broadband Radio Access Networks (BRAN), High Performance Radio Local Area Network (HIPERLAN) Type 2, System Overview, ETSI TR 101 683 version 1.1.1, 2000.

165. European Telecommunications Standards Institute, Broadband Radio Access Networks (BRAN), High Performance Radio Local Area Network (HIPERLAN) Type 2, Physical (PHY) Layer, ETSI TR 101 475 version 1.3.1, 2001.

166. European Telecommunications Standards Institute (ETSI), Broadband Radio Access Networks (BRAN), High Performance Radio Local Area Network (HIPERLAN) Type 2, Data Link Control (DLC) Layer; Part 1: Basic Data Transport Functions, ETSI TS 101 761-1 version 1.3.1, 2001.

167. European Telecommunications Standards Institute (ETSI), Broadband Radio Access Networks (BRAN), High Performance Radio Local Area Network (HIPERLAN) Type 2, Data Link Control (DLC) Layer; Part 2: Radio Link Control (RLS) sublayer, ETSI TS 101 761-2 version 1.3.1, 2002.

168. J. W. Craig, A New, Simple, and Exact Result for Calculating the Probability of Error for Two-Dimensional Signal Constellations, *IEEE MILCOM 1991*, November 1991.

169. M. K. Simon and M.-S. Alouini, *Digital Communication over Fading Channels*, 2nd edition, Wiley, 2005.

170. Bluetooth SIG, *Specification of the Bluetooth System*, vol. 1, Corem version 1.1, 2001.

171. B. Ghimire, G. Auer, and H. Haas, Busy Bursts for Trading-off Throughput and Fairness in Cellular OFDMA-TDD, *EURASIP Journal on Wireless Communications and Networking*, Article ID 462396, vol. 2009, 2009.

172. Q. Jing and Z. Zheng, Distributed Resource Allocation Based on Game Theory in Multi-cell OFDMA Systems, *International Journal of Wireless Information Networks*, 16, 44–50, 2009.

173. D. Monderer and L. Shapley, Potential Games, in *Games and Economic Behavior*, vol. 14, 1996, pp. 124–143.

174. H. Harada and R. Prasad, *Simulation and Software Radio for Mobile Communications*, Artech House, 2002.

175. D. Lopez-Perez, A. Valcarce, G. De La Roche, E. Liu, and J. Zhang, Access Methods to WiMAX Femtocells: A Downlink System-Level Case Study, in *11th IEEE Singapore International Conference on Communication Systems (ICCS 2008)*, November 2008, pp. 1657–1662.

176. H. Claussen, Performance of Macro- and Co-Channel Femtocells in a Hierarchical Cell Structure, in *IEEE PIMRC 2007*, September 2007.

177. L. T. W. Ho and H. Claussen, Effects of User-Deployed, Co-Channel Femtocells on the Call Drop Probability in a Residential Scenario, in *IEEE PIMRC 2007*, September 2007.

178. P. Xia, V. Chandrasekhar, and J. Andrews, Open vs. Closed Access Femtocells in the Uplink, *IEEE Transactions on Wireless Communications*, 9(12), 3798–3809, December 2010.

179. V. Chandrasekhar, M. Kountouris, and J. G. Andrews, Coverage in Multi-Antenna Two-Tier Networks, *IEEE Transactions on Wireless Communications*, 8(10), 5314–5327, 2009.

180. A. Valcarce, D. López-Pérez, G. De La Roche, and J. Zhang, Limited Access to OFDMA Femtocells, *IEEE PIMRC 2009*, September 2009.

181. H.-S. Jo, C. Mun, J. Moon, and J.-G. Yook, Self-Optimized Coverage Coordination in Femtocell Networks, *IEEE Transactions on Wireless Communications*, 9(10), 2977–2982, 2010.

182. D. Lopez-Perez, A. Ladanyi, A. Juttner, and J. Zhang, OFDMA Femtocells: A Self-Organizing Approach for Frequency Assignment, in *IEEE PIMRC 2009*, September 2009.

183. C.-Y. Oh, M. Y. Chung, H. Choo, and T.-J. Lee, A Novel Frequency Planning for Femtocells in OFDMA-Based Cellular Networks Using Fractional Frequency Reuse, in *International Conference on Computational Science and Its Applications ICCSA*, March 2010, pp. 96–106.

184. I. Guvenc, M.-R. Jeong, F. Watanabe, and H. Inamura, A Hybrid Frequency Assignment for Femtocells and Coverage Area Analysis for Co-Channel Operation, *IEEE Communications Letters*, 12, 880–882, 2008.

185. H.-C. Lee, D.-C. Oh, and Y.-H. Lee, Mitigation of Inter-Femtocell Interference with Adaptive Fractional Frequency Reuse, in *IEEE ICC 2010*, May 2010.

186. WiMAX Forum, Fixed, Nomadic, Portable and Mobile Applications for 802.16-2004 and 802.16e WiMAX Networks, White Paper, WiMAX Forum, November 2005.

187. IEEE 802.16e, IEEE Standard for Local and Metropolitan Area Networks, Air Interface for Fixed Broadband Wireless Access Systems, Amendment 2: Physical and Medium Access Control Layers for Combined Fixed and Mobile Operation in Licensed Bands and Corrigendum 1, 2005.

188. L. Nuaymi, *WiMAX: Technology for Broadband Wireless Access*, Wiley, 2007.

189. H.-Y. Wei, S. Ganguly, R. Izmailov, and Z. J. Haas, Interference-Aware IEEE 802.16 WiMax Mesh Networks, *IEEE VTC-Spring 2005*, May 2005.

190. Y. Zhang, H. Hu, and H.-H. Chen, QoS Differentiation for IEEE 802.16 WiMAX Mesh Networking, *Mobile Networks and Applications*, 13(1–2), 19–37, 2008.

191. B. Makarevitch, Adaptive Resource Allocation for WiMAX, *IEEE PIMRC 2007*, September 2007.

192. L. Jorguseski, T. M. H. Le, E. R. Fledderus, and R. Prasad, Downlink Resource Allocation for Evolved UTRAN and WiMAX Cellular Systems, *IEEE PIMRC 2008*, September 2008.

193. K. Etemad, Overview of Mobile WiMAX Technology and Evolution, *IEEE Communications Magazine*, 46(10), 31–40, 2008.

194. S. Ahmadi, An Overview of Next-Generation Mobile WiMAX Technology, *IEEE Communications Magazine*, 47(6), 84–98, 2009.

195. WiMAX Forum, Mobile WiMAX Part 1: A Technical Overview and Performance Evaluation, August 2006.

196. B.P. Tiwari, Enabling Reuse 1 in 4G Networks, www.beyond4g.org, July 2010.

197. WiMAX Forum, WiMAX: A Way Forward in India, August 2010.

198. H. Ekström, A. Furuskär, J. Karlsson, M. Meyer, S. Parkvall, J. Torsner, and M. Wahlqvist, Technical Solutions for the 3G Long-Term Evolution, *IEEE Communications Magazine*, 44(3), 38–45, 2006.

199. H. G. Myung and D. J. Goodman, *Single Carrier FDMA: A New Air Interface for Long Term Evolution*, Wiley, 2008.

200. H.G. Myung, J. Lim, and D.J. Goodman, Single Carrier FDMA for Uplink Wireless Transmission, *IEEE Vehicular Technology Magazine* 48(1), 30–38, 2006.

201. Motorola, Long Term Evolution (LTE): Overview of LTE Air–Interface, Technical white paper, 2007.

202. 3rd Generation Partnership Project (3GPP), 3GPP TR 25.892 Feasibility Study for Orthogonal Frequency Division Multiplexing (OFDM) for UTRAN Enhancements, v6.0.0, June 2004.

203. D. Astély, E. Dahlman, A. Furuskär, Y. Jading, M. Lindström, and S. Parkvall, LTE: The Evolution of Mobile Broadband, *IEEE Communications Magazine*, 47(4), 44–51, 2009.

204. 3rd Generation Partnership Project (3GPP), 3GPP TR 36.913 3GPP TSG RAN Requirements for Further Advancements for Evolved Universal Terrestrial Radio Access (E-UTRA) (LTE-Advanced), version 8.0.1, Release 8, 2008.

205. H. Wu and T. Haustein, Energy and Spectrum Efficient Transmission Modes for the 3GPP-LTE Uplink, *IEEE PIMRC 2007*, September 2007.

206. F. D. Calabrese, P. H. Michaelsen, C. Rosa, M. Anas, C. Úbeda Castellanos, and D. López Villa, Search-Tree Based Uplink Channel Aware Packet Scheduling for UTRAN LTE, *IEEE VTC-Spring 2008*, May 2008.

207. I. C. Wong, O. Oteri, and W. McCoy, Optimal Resource Allocation in Uplink SC-FDMA Systems, *IEEE Transactions on Wireless Communications*, 8(5), 2161–2165, 2009.

208. G. Berardinelli, L.A. Ruiz de Temino, S. Frattasi, M. Rahman, and P. Mogensen, OFDMA vs. SC-FDMA: Performance Comparison in Local Area IMT-A Scenarios, *IEEE Wireless Communications*, 15(5), 64–72, 2008.

209. J. Ellenbeck, H. Al-Shatri, and C. Hartmann, Performance of Decentralized Interference Coordination in the LTE Uplink, *IEEE VTC-Fall 2009*, September 2009.

210. A. Gotsis, D. Komnakos, and P. Constantinou, Dynamic Subchannel and Slot Allocation for OFDMA Networks Supporting Mixed Traffic: Upper Bound and a Heuristic Algorithm, *IEEE Communications Letters*, 13(8), 576–578, 2009.

211. T.E. Kolding, F. Frederiksen, and A. Pokhariyal, Low-Bandwidth Channel Quality Indication for OFDMA Frequency Domain Packet Scheduling, *IEEE ISWCS 2006*, September 2006.

212. S.-B. Lee, I. Pefkianakis, A. Meyerson, S. Xu, and S. Lu, Proportional Fair Frequency-Domain Packet Scheduling for 3GPP LTE Uplink, *IEEE INFOCOM 2009*, April 2009.

213. S. Venkatesan, A. Lozano, and R. Valenzuela, Network MIMO: Overcoming Intercell Interference in Indoor Wireless Systems, *Asilomar Conference on Signals, Systems and Computers*, November 2007.

214. W. Saad, Z. Han, M. Debbah, A. Hjørungnes, and T. Basar, Coalitional Game Theory for Communication Networks: A Tutorial, *IEEE Signal Processing Magazine, Special Issue on Game Theory*, 26(5), 77–97, 2009.

215. W. Saad, Z. Han, M. Debbah and A. Hjørungnes, A Distributed Coalition Formation Framework for Fair User Cooperation in Wireless Networks, *IEEE Transactions on Wireless Communications*, 8(9), 4580–4593, 2009.

216. W. Saad, Z. Han, T. Basar, M. Debbah and A. Hjørungnes, Coalitional Games for Distributed Eavesdroppers Cooperation in Wireless Networks, *Third International Workshop on Game Theory in Communication Networks (Gamecomm)*, October 2009.

217. W. Saad, A. Hjørungnes, Z. Han and T. Basar, Network Formation Games for Wireless Multi-hop Networks in the Presence of Eavesdroppers (invited paper), *Third International Workshop on Computational Advances in Multi-Sensor Adaptive Processing (CAMSAP)*, December 2009.

218. W. Saad, Q. Zhu, T. Basar, Z. Han and A. Hjørungnes, Hierarchical Network Formation Games in the Uplink of Multi-hop Wireless Networks, *IEEE GlobeCom 2009*, December 2009.

219. W. Saad, Z. Han, T. Basar, M. Debbah and A. Hjørungnes, Physical Layer Security: Coalitional Games for Distributed Cooperation, *International Symposium on Modeling and Optimization in Mobile, Ad Hoc and Wireless Networks (WiOpt)*, June 2009.

220. J. T. Louhi, Energy Efficiency of Modern Cellular Base Stations, *International Telecommunications Energy Conference (INTELEC)*, September 2007.

221. L. M. Correia, D. Zeller, O. Blume, D. Ferling, Y. Jading, I. G. Gunther Auer, and L. Van der Perre, Challenges and Enabling Technologies for Energy Aware Mobile Radio Networks, *IEEE Communications Magazine*, 48(11), 66–72, 2010.

222. EARTH, Energy Aware Radio and neTwork tecHnologies, https://www.ict-earth.eu, EU Funded Research Project FP7-ICT-2009-4-24733-EARTH, Jan. 2010–June 2012.

223. M. A. Marsan, L. Chiaraviglio, D. Ciullo, and M. Meo, Optimal Energy Savings in Cellular Access Networks, *IEEE ICC Workshops 2009*, June 2009.

224. D. Ciullo, L. Chiaraviglio, M. Meo, and M. A. Marsan, Energy-Aware UMTS Access Networks, *International Symposium on Wireless Personal Multimedia Communications (WPMC'08)*, September 2008.

225. S. Zhou, J. Gong, Z. Yang, Z. Niu, and P. Yang, Green Mobile Access Network with Dynamic Base Station Energy Saving, *MobiCom 2009*, September 2009.

226. Z. Niu, Y. Wu, J. Gong, and Z. Yang, Cell Zooming for Cost-Efficient Green Cellular Networks, *IEEE Communications Magazine*, 48(11), 74–79, 2010.

227. E. Oh and B. Krishnamachari, Energy Savings through Dynamic Base Station Switching in Cellular Wireless Access Networks, *IEEE Globecom 2010*, December 2010.

228. D. Willkomm, S. Machiraju, J. Bolot, and A. Wolisz, Primary Users in Cellular Networks: A Large-Scale Measurement Study, *IEEE DySPAN*, October 2008.

229. NEC Corporation, Self Organizing Network NEC's proposals for next-generation radio network management, white paper, February 2009.

230. D. Niyato and E. Hossain, Queue-Aware Uplink Bandwidth Allocation and Rate Control for Polling Service in IEEE 802.16 Broadband Wireless Networks, *IEEE Transactions on Mobile Computing*, 5(6), 668–679, 2006.

231. S. Kim and J.-W. Lee, Joint Resource Allocation for Uplink and Downlink in Wireless Networks: A Case Study with User-Level Utility Functions, *IEEE VTC-Spring 2009*, April 2009.

232. J. Price and T. Javidi, Leveraging Downlink for Efficient Uplink Allocation in a Single-Hop Wireless Network, *IEEE Transactions on Information Theory*, 53(11), 4330–4339, 2007.

233. M. Kou and Y. Zhen, Dynamic Uplink/Downlink Resource Allocation For TDD OFDMA Access Network, *International Conference on Communications and Mobile Computing*, January 2009.

234. 3rd Generation Partnership Project (3GPP), 3GPP TS 23.060 General Packet Radio Service (GPRS) Service Description, version 8.4.0, Release 8, 2008.

235. 3rd Generation Partnership Project (3GPP), 3GPP TS 23.401 General Packet Radio Service (GPRS) Enhancements for Evolved Universal Terrestrial Radio Access Network (E-UTRAN) Access, version 8.4.1, Release 8, 2008.

236. D. Niyato and E. Hossain, Integration of WiMAX and WiFi: Optimal Pricing for Bandwidth Sharing, *IEEE Communications Magazine*, 45(5), 140–146, 2007.

237. O. Ognenoski, V. Rakovic, M. Bogatinovski, V. Atanasovski, and L. Gavrilovska, User Perception of QoS and Economics for a WiMAX Network in a Backup Scenario, *1st Wireless Communication, Vehicular Technology, Information Theory and Aerospace and Electronic Systems Technology (Wireless VITAE) International Conference*, May 2009, pp. 102–106.

238. L. Le and E. Hossain, Resource Allocation for Spectrum Underlay in Cognitive Radio Networks, *IEEE Transactions on Wireless Communications*, 7(12), 5306–5315, 2008.

239. A. A. El-Saleh, M. Ismail, M. A. M. Ali, and A. N. H. Alnuaimy, Capacity Optimization for Local and Cooperative Spectrum Sensing in Cognitive Radio Networks, *International Journal of Electronics, Circuits and Systems*, 3(3), 132–138, 2009.

240. B.-J. Kang, J.-W. Seo, and S.-W. Ban, Performance Analysis of Spectrum Sensing for RF Receiver Structure in Cognitive Radio Networks, *Proceedings of the 8th WSEAS International Conference on Applied Electromagnetics, Wireless and Optical Communications*, March 2010.

241. S. Srinivasa and S. A. Jafar, The Throughput Potential of Cognitive Radio: A Theoretical Perspective, *Asilomar Conference on Signals, Systems and Computers (ACSSC 2006)*, 2006.

242. V. D. Chakravarthy, Z. Wu, A. Shaw, M. A. Temple, R. Kannan, and F. Garber, A General Overlay/Underlay Analytic Expression Representing Cognitive Radio Waveform, *International Waveform Diversity and Design Conference*, June 2007.

243. S. Sengupta, M. Chatterjee, and R. Chandramouli, Dynamic Pricing for Service Provisioning and Network Selection in Heterogeneous Networks, *Physical Communication*, 2(1–2), 138–150, 2009.

244. S. J. Shellhammer, Spectrum Sensing in IEEE 802.22, *IAPR Workshop on Cognitive Information Processing*, June 2008.

245. 3rd Generation Partnership Project (3GPP), 3GPP TS 36.101 3GPP TSG RAN Evolved Universal Terrestrial Radio Access (E-UTRA) User Equipment (UE) Radio Transmission and Reception, version 9.4.0, Release 9, 2010.

246. E. Yaacoub and Z. Dawy, Enhancing the Performance of OFDMA Underlay Cognitive Radio Networks via Secondary Pattern Nulling and Primary Beam Steering, *IEEE WCNC 2011*, March 2011.

INDEX

Resource Allocation in Uplink OFDMA Wireless Systems: Optimal Solutions and Practical Implementations,
Elias E. Yaacoub and Zaher Dawy.
© 2012 by the Institute of Electrical and Electronics Engineers, Inc. Published 2012 by John Wiley & Sons, Inc.